Neuroscience
and the Law

*Brain, Mind, and
the Scales of Justice*

A REPORT ON
AN INVITATIONAL MEETING CONVENED BY
THE AMERICAN ASSOCIATION FOR THE
ADVANCEMENT OF SCIENCE
and
THE DANA FOUNDATION

Edited by
Brent Garland

D
**DANA
PRESS**

AAAS
ADVANCING SCIENCE, SERVING SOCIETY

New York | Washington, D.C.   Washington, D.C.

The Charles A. Dana Foundation
745 Fifth Avenue, Suite 900
New York, New York 10151

Published by
Dana Press
New York | Washington, DC

And

American Association for the Advancement of Science
1200 New York Ave., NW
Washington, DC 20005

ISBN (paper): 1-932594-04-3

Library of Congress Cataloging-in-Publication Data

Neuroscience and the law : brain, mind, and the scales of justice : a report on an invitational meeting convened by the American Association for the Advancement of Science and the Dana Foundation / edited by Brent Garland; foreword by Mark S. Frankel.
    p. cm.
  Includes index.
  ISBN 1-932594-04-3 (pbk. : alk. paper)
  1. Neuroscientists--Legal status, laws, etc.--United States. 2. Neurosciences--Research. 3. Brain--Imaging. I. Garland, Brent. II. American Association for the Advancement of Science. III. Charles A. Dana Foundation.
  KF2910.N45N48 2004
  153--dc22
                              2004002405

Pg. 56: From Libet, 1999, *Journal of Consciousness Studies*, 6 (8-9), pg. 50.
Used by permission of Imprint Academic.
Pg. 64: From *Freedom Evolves* by Daniel C. Dennett, copyright 2003, by Daniel C. Dennett. Used by permission of Viking Penguin, a division of Penguin Group (USA) Inc.

Design by Potter Publishing Studio
Printed by Action Printing, USA

This report reflects the effort of the editor to provide an accurate account of the content of the invitational meeting. The report does not necessarily represent the views of the American Association for the Advancement of Science or the Dana Foundation.

*I*n addition to this detailed report, a free, 30-page summary report is available for a general audience that wishes to become more familiar with the issues examined by this project. If you are interested in obtaining the summary report, please contact:

Randy Talley
Dana Press
900 15th Street, NW
Washington, DC 20005

———

E-mail: rtalley@dana.org

———

Fax: 202-408-5599

Contents

Foreword

By Mark S. Frankel

I T IS COMMONPLACE to think of human beings as rational creatures, and at the seat of our rationality is a three-pound organ, the brain. Although at present we know very little about the interactions between the brain and the mind, advances in neuroscience are beginning to reveal more and more about the underlying processes that shape cognition and emotion. This report examines new research and tools for investigating the functioning human brain, as well as some potential advances on the horizon. Current neuroscience developments hold great promise for improving our understanding of disease and behavior and eventually reducing human suffering. But they also carry the danger that they will be misused in ways that may thwart human potential, unfairly deny benefits to those in need, or threaten long-standing legal rights.

As noted in the report, improving the ability to confirm truth-telling (that is, lie detection) or to enhance memory through noninvasive procedures will present challenges to courts' treatment of witnesses. Should the courts be permitted to compel a witness in litigation to submit to such procedures, and, if so, under what circumstances? If the procedures are made available voluntarily, should jurors be told of a witness's willingness or unwillingness to undergo such a procedure? Would increased ability to predict potential violent tendencies or intellectual achievement lead to discrimination and denial of opportunities to those who do not meet certain standards? These are just a few of the intriguing questions covered in the report.

The American Association for the Advancement of Science (AAAS) is the world's largest multidisciplinary scientific society and publisher of one of the preeminent peer-reviewed journals, *Science*. Since 1974 the association has had a joint committee with the American Bar Association—the National Conference of Lawyers and Scientists—that focuses on emerging public policy issues at the nexus of science and law. Staff in the AAAS Program on Scientific Freedom, Responsibility and Law developed this project on neuroscience and the law in cooperation with members of the National Conference of Lawyers and Scientists. The Dana Foundation was approached for funding for the project because of its long-standing commitment to neuroscience research and its more recent interest in exploring the broader social implications of such research. It has proved to be a very productive collaboration.

Predicting the future is always a challenge, since one must speculate both on the direction that advances in science and technology will take and on what the effects of such innovations will be. Nevertheless, we believe it serves both science and the law to take on that task with regard to neuroscience, albeit cautiously, to promote a broader and more robust public dialogue. The range of issues covered is formidable, including such matters as free will and responsibility, enhancement of cognitive functions, prediction of behavior, and determination of truth or falsehoods.

The analysis of these issues is in general conducted within the framework of the U.S. legal system. The relationship between science, technology, and the law is a dynamic one, with each exerting influence over the other. Advances in science and technology can, and often do, challenge traditional legal concepts and practices. For example, the use of DNA technology in the criminal justice system has significantly changed practices in law enforcement investigations and evidentiary proceedings. It is also the case that the fate of science and technology is shaped, in part, by how the law is interpreted and applied. It makes sense, therefore, to engage in an effort to highlight the special issues arising from the intersection of neuroscience and the law.

This volume is a report based on an invitational workshop convened by AAAS and the Dana Foundation at the Dana Center in Washington, D.C., on September 12–13, 2003. The workshop was organized around four commissioned papers, two by neuroscientists and two by legal scholars, which were distributed in advance to all workshop invitees. Participating in the meeting were leaders from several fields of neuroscience research, as well as legal scholars and attorneys and state and federal judges. For one and a half days, the 27 participants (a list of whom can be found at the end of this volume) wrestled with the formidable task of projecting developments to which neuroscience might lead us and how the law might affect those developments and be affected by them. Participants were specifically asked to distinguish, where possible, between near-term advances and those that were further downstream.

Part I of this book, produced by AAAS staff, distills those deliberations, and Part II presents all four commissioned papers, revised by their authors following the workshop. The results of this effort, of course, are constrained by the knowledge, experience, and foresight of the people involved. They made no pretense of being engaged in producing the final word on the matter. Through their many hours of discussion, however, they did make remarkable progress in helping to map the terrain and, we hope, fuel the work of those who will choose to build on this initial endeavor.

In recent years, it has become strikingly apparent how science has transformed our lives in ways that are often unpredictable. Awareness of that gives those of us involved in this project a deep sense of humility in attempting to discern a course for neuroscience research and technology. But the promise of neuroscience is so great that it seems wise to consider how best to advance such research in a manner consistent with broader social values. Science cannot function in isolation from the larger community that feels its effects, whether perceived as good or bad. The decisions to pursue research and to develop the public policy associated with it must be accessible to broad participation so that the values of all stakeholders can be carefully considered and weighed. We hope this report and its thought-provoking papers will be a major first step in that direction.

Acknowledgments

AAS CONVENED A GROUP of neuroscientists, legal scholars, lawyers, and judges in order to begin to limn the issues that arise at the interface between neuroscience and the law. They met the challenge admirably and, in doing so, earned our gratitude. Everyone involved in designing and organizing the meeting is deeply indebted to the commitment and contributions made by the participants. We hope the report captures most of the insights, observations, and wise counsel they provided.

From the initial stages of this project until its end, we have benefited from the support of Floyd Bloom, immediate past Chairman of the AAAS Board of Directors, and Alan I. Leshner, CEO of AAAS. We are grateful for their help.

As the editor of this report, I am deeply indebted to my AAAS colleagues Mark Frankel and Kristina Schaefer, and wish to thank them for the key intellectual roles they played in developing both the meeting and this report. Mark Frankel, Director of the Program on Scientific Freedom, Responsibility and Law at AAAS, helps turn mountains into molehills through his guidance and support. His willingness to put his vast experience at my disposal is an incalculable gift. His role in developing and conceiving the project, and in preparing this report, was crucial. Likewise, I am grateful to Kristina Schaefer, whose efforts in the conception, coordination, and execution of this workshop, in addition to her significant contributions to the meeting itself and this volume, were indispensable.

The Dana Foundation's support of brain research, and its willingness to reach beyond its traditional areas of support to fund this meeting, speaks volumes about its commitment to advancing society's knowledge of the brain and understanding of the potential impact of that knowledge on our lives. The foundation provided not only financial support but also the venue in which the meeting was convened. In addition, many Dana staff members offered logistical planning support and assistance, for which we are thankful. Particular acknowledgment is due to Barbara Gill and Karen Graham.

Equally supportive and helpful were the staff of the Dana Press, and we would specifically like to thank Leticia Barnes, Jane Nevins, and Randy Talley, who have guided us in the production of this volume.

The authors of the commissioned papers laid the intellectual framework for this discussion. We are very grateful to Mike Gazzaniga and Megan Steven, Hank Greely, Stephen Morse, and Larry Tancredi for their invaluable contributions.

We thank those who volunteered to read an earlier draft of this report. The comments and feedback of Erica Beecher-Monas, Owen Jones, Chuck O'Brien, Don Pfaff, Haskell Pitluck, Adina Roskies, and Larry Tancredi were of great assistance.

The work that led up to the meeting, and the work that followed, involved many current and former staff members of AAAS. Thanks to Christine Bellorde, Melanie Frankel, Dana Greenspon, Hilary Leeds, Miguel Prietto, and Deborah Runkle. The efforts of Crystal Liu and Bryn Lander in compiling our glossary merit a special thanks.

—*Brent Garland*

PART I

Neuroscience and the Law

A Report

by Brent Garland

Framing the Issues

Neuroscience really, over the last 30 years, has just blossomed at every point, and each year brings a greater understanding of the mechanical way with which we perceive, we remember, we speak, we feel . . . It is that sense of understanding the brain that really brings us here today with force, and that [sense of understanding] is the one that we have to come to grips with.

—Participant scientist at the meeting

IT HAS BEEN 14 YEARS since President George H. W. Bush proclaimed the 1990s to be the Decade of the Brain,[1] and now, in the new millennium, knowledge and applications resulting from increased emphasis on brain science are beginning to emerge. While advances in neuroscience continue at a rapid rate, their ethical and legal implications are only beginning to be considered. A recent article in the *Economist* made the point that the link between brain and behavior is much closer than the link between genes and behavior, yet the public debate about genetics research and its broad social implications far outweighs that given to neuroscience and technology.[2]

Even at the level of a "first glance," neuroscience raises numerous issues with respect to some of the core constructs of the law, such as competency, free will, and the genesis of violent behavior. For example,

---

[1] President George H. W. Bush, Proclamation, "Decade of the Brain, 1990–2000, Proclaimation 6158," *Federal Register* 55, n. 140 (1990): 29553.

[2] "Open Your Mind," *Economist*, May 23, 2002.

the discovery of a neurological predisposition to violence would pose a host of controversial questions, including whether "preemptive" treatment is desirable, whether we might "mark" the person for increased surveillance by authorities, whether there would be an increased risk for discrimination against such a person, and whether substantial changes might occur in how society approaches the punishment and rehabilitation of criminals.

The question of how developments in neuroscience might interact with the law led AAAS and the Dana Foundation to convene a meeting with members drawn from both the legal and neuroscience communities. For a day and a half, lawyers, judges, law professors, philosophers, psychologists, psychiatrists, and neuroscientists engaged in a conversation focused on the relationship between neuroscience and law and sought to contribute to the larger public discourse by identifying some central issues and suggesting directions for future efforts.

The 27 meeting participants discussed a broad range of topics, informed in part by the four commissioned papers that make up Part II of this book. Here, in Part I, we distill the key ideas and concerns that arose from the discussion. Each participant from neuroscience or the law entered the conversation as an expert in his or her own discipline. The group's task was to compare concerns in both fields, identify issues of interest, and begin to develop a way of thinking about them. This report does not attempt to pronounce a definitive set of findings in an area that is in great flux, but is instead an account of the issues and considerations that focused the meeting's firepower during the hours of discussion, along with some suggestions for further consideration.

The four papers anchoring the dialogue and reprinted in Part II were written by three neuroscientists and two legal scholars, prominent in their respective fields. In "Free Will in the Twenty-first Century: A Discussion of Neuroscience and the Law," coauthors Michael Gazzaniga[3] and Megan Steven[4] present the philosophical arguments and neuroscientific

---

[3] Dr. Gazzaniga is a psychologist, Director of the Cognitive Neuroscience Program, and Dean of the Faculty at Dartmouth College.

[4] Ms. Steven is a doctoral candidate in medical sciences at the University of Oxford in England.

findings addressing free will and legal responsibility. "Neuroscience Developments and the Law," by Laurence Tancredi,[5] focuses primarily on issues of cognition and looks in detail at issues of competency, brain death, cognitive enhancement, and lie detection. "Prediction, Litigation, Privacy, and Property: Some Possible Legal and Social Implications of Advances in Neuroscience," by Henry Greely,[6] addresses the prediction of behavior, privacy concerns, patent and intellectual property issues, and uses of neuroscience in litigation. "New Neuroscience, Old Problems," by Stephen Morse,[7] examines a variety of issues, including free will and responsibility, the desirability and permissibility of enhancements of cognitive abilities that are within the normal range, informed consent, and the development of legal doctrine generally, and concludes with a brief examination of the admissibility of new neuroscientific evidence.

The first problem in assessing the weight of neuroscience discoveries on the scales of justice is to answer the question of how to approach the relevant issues. A few fundamental points need to be considered.

First, *is neuroscience likely to actually impact the law?* Developments in neuroscience may well have substantial impact on how the law views people and behavior, but the legal system should be able to assimilate and use even revolutionary science without upending its own fundamental structure. The legal system is generally robust and has withstood civil war, social strife, and a variety of other substantial challenges to existing law and legal assumptions. Therefore, the focus should be on a realistic assessment of the advances in neuroscience and their potential for good or ill effects in law, with less concern about paradigm-altering discoveries. (Also considered at the meeting, though to a much lesser extent, was how the law might impact neuroscience, which is briefly addressed in the report as well.)

---

[5] Dr. Tancredi is a psychiatrist, lawyer, and Clinical Professor of Psychiatry at New York University School of Medicine.

[6] Professor Greely is a lawyer, Deane F. and Kate Edelman Johnson Professor of Law at Stanford School of Law, Professor (by courtesy) of Genetics, Director of the Center for Law and the Biosciences, and Co-Director of the Program in Genomics, Ethics, and Society at Stanford University.

[7] Dr. Morse is a psychologist, lawyer, and Ferdinand Wakeman Hubbell Professor of Law at the University of Pennsylvania School of Law.

Second, *how far ahead is it reasonable to look in trying to foresee discoveries and their legal implications?* Much of the science is still in its nascent stages. A few areas are more fully developed and are more immediately relevant, and those areas are noted later in this report.

Finally, given the breadth of topics addressed by both neuroscience and the law, *what analytical framework for considering neuroscience developments in relation to the law might be helpful?* Since neuroscience encompasses many fields (psychology, psychiatry, and psychopharmacology, among others) and addresses such far-ranging topics as how we perceive, speak, feel, and think, one of the most difficult aspects of addressing the topics is merely to find an organizing principle to discuss the science. As one participant noted, modern neuroscience provided "a host of findings, all over the place."

The simplest approach is a two-prong classification. In one class are neuroscience findings and technologies related to monitoring and imaging the brain. Generally, "monitoring and imaging" includes the prediction of behavior, lie detection, and brain death, among other concepts. The other class consists of manipulations of the human brain and includes enhancement and pharmacological treatment of addiction. This distinction is the organizing principle used in this report. Additionally, from a legal perspective, a handful of general, crosscutting legal issues do not fit easily into this framework. Accordingly, they are explored separately.

Monitoring and Imaging the Brain

BROAD RANGE OF ISSUES, from brain death to the prediction of behavior, are encompassed in the consideration of imaging and monitoring the brain. As imaging technologies continue to improve, neuroscientists are taking increasingly fine-grained pictures of brain function, producing an ever better sense of what "happens" in our brains as we perform tasks, experience emotion, and engage in various behaviors. While such data may benefit us by helping to explain how and even why we act in certain ways, we should be sensitive to the concomitant risk that we will misuse such knowledge or rely too much on deterministic explanations.

Prediction of Behavior

If a single topic captures the sense of promise and risk, the ability neuroscience may provide for the prediction of behavior (and capacity) is it. As one scientist at the meeting put it:

> [It] is virtually inevitable that science is going to give us the ability to make, with a certain amount of confidence, [and] higher than we now have, predictions about how people are going to behave. And the question that's going to come, like it or not, is . . . "What do you do with that information?"

The perspective of many neuroscientists is that a descriptive biology of behavior will soon be available (and is likely to incorporate both genetic and neuroscientific knowledge). But a predictive biology is not

7

on the near horizon. It is apparent, however, that the optimal time to begin the dialogue about the appropriate uses of neuroscience to predict behavior is before the science is fully developed. A number of scientific and technological discoveries—from the splitting of the atom to the development of cloning—have demonstrated that when we don't think about potential social and ethical implications of technologies before they are fully developed, we often feel overwhelmed and unprepared for their use. Similarly, where the science may have powerful and profound effects, preparation for its potential uses may prevent possible abuses. While the potential uses and misuses of predictive knowledge are evident, the first issue to consider is what it means to predict behavior based on brain function and structure—that is, what about free will and personal responsibility?

Blame It on the Brain?

When one begins to talk about predicting behavior based on brain imaging, the discussion often turns toward the question of the relationships between neurological determinism, free will, and legal responsibility. While all the commissioned papers touch on this issue to some degree, both Gazzaniga and Morse specifically address it in some depth, and readers with particular interest in the philosophical underpinnings of the concept of free will should refer to those papers for a broader discussion of the issue.[8]

The short version of the issue is this: as neuroscience reveals more about the brain, it becomes increasingly clear that the brain is a physical entity governed by the principles and rules of the physical world. In addition, it is increasingly clear that the brain determines the mind. If the physical world is determined, in the sense that the principles and rules of the physical world

---

[8] There are a variety of philosophical positions on the existence of free will, and on the concepts of dualism and determinism. For the purposes of this report, and in line with the nature of the discussions at the meeting, the deterministic nature of the brain is taken as given, and the issue of dualism is decided in favor of materialism. Some other positions covered in Gazzaniga's and Morse's papers are philosophically interesting and very complex but were not extensively discussed by participants and so are not included in the body of this report. See Gazzaniga, Morse, this volume.

allow us to predict with great accuracy what will happen in the physical world (assuming we know the starting conditions and other relevant data), then the brain, too, must be determined. It follows that if the brain is determined, the mind must also be determined.

This leads us to the question that one neuroscientist participant framed so well:

> If our brains are determined and if the brain is the necessary and sufficient organ that enables the mind, then we are left with the question: Are the thoughts that arise from our mind also determined? Is the free will we seem to experience just an illusion, and if the free will is an illusion, must we revise our concept of what it means to be personally responsible for our actions?

Or more bluntly, if we are mere mechanisms, controlled by our mechanistic brains, then how can we have free will? And if we do not have free will, then how can we be held responsible for our own action, whether that consists of signing a contract or committing a murder?

The question of the existence of free will is one that philosophers have contemplated for centuries as central to humankind's purpose and meaning. In addition, the notion of responsibility (arising from our having free will) is central to our legal system—we hold a person responsible for her actions only to the extent that she was free to act (as reflected in the defenses of coercion, insanity, and infancy, which absolve or limit one's responsibility for one's actions, as they were not truly free). However, as the debate is currently being infused with new information from neuroscience about the mechanistic nature of our brains, the question of free will may be reconsidered from a different perspective.

While some concerns about the challenges to free will and personal responsibility deserve attention, neuroscience is ultimately very unlikely to result in the overturning of those concepts, at least in the context of the law. Responsibility is a human (or societal) construct, and not one that will be measured by a brain scan. Gazzaniga and Morse, among others, present a variety of arguments to support this idea, buttressing the view that neuroscience is unlikely to undo our concept of free will.

One such argument offered by Gazzaniga emphasizes the social construction of responsibility, which is rooted in observations regarding the ability of simple rule-based deterministic systems to learn new rules and behaviors. We, as a society, expect individuals to learn those rules, and for there to be consequences for those individuals who do not. The rules we construct (the "rules of engagement"), in Gazzaniga's view, are ones human beings can learn, and they form a social construct that provides "kind of . . . all the free will we want."

One scientist provided a perspective on the social construction of responsibility that meshes well with Gazzaniga's argument, but sees the very biology of the brain as having a role:

> We are responsible because our brains tell us we feel we must be. We are creatures of rules. I think that comes out because our brains have evolved the way they have. We feel there should be responsibility. Again for biological reasons, we feel that if one person injures another, there should be some sort of payback. This is . . . a biological reality, and in fact, I would argue . . . that our brains determine that we are creatures of rules . . . and that the view of man as a rational creature is a construct that lets us put into place the rules that . . . we feel are necessary.

Another of Gazzaniga's arguments is based on the gap in time—on the order of milliseconds—between when our senses receive information and when it becomes part of our conscious experience. Or, as he puts it, "this sort of realization that there is time, there is brewing time going on . . . before you are consciously aware of it." The idea is that in that gap, a person can choose not to act, and that is when free will may play a role. In his paper, Gazzaniga quotes Vilayanur Ramachandran: "our conscious minds may not have free will but rather 'free won't!'"[9] The gap allows a time for intervention, for choosing a different course of action, even if it happens outside consciousness.

The approach offered by Morse suggests that we understand ourselves to be rational creatures, not cold and calculatingly rational but moved and motivated by reason. Morse argues that, because of the way

---

[9] See Gazzaniga, this volume.

we are constructed, there is no way for human beings to get outside our perception of ourselves as rational beings. We will therefore continue to behave and interact as if we are rational, and to base decisions about responsibility on our beliefs and perceptions concerning rationality, even if we are completely determined. Accordingly, our laws and rules will reflect our understanding of ourselves as rational actors.

While Morse uses a simple, commonsense definition of rationality,[10] several participants made the point that rationality is not well defined or understood, and that neuroscience could play a role in elucidating the construct of rationality itself. Morse agreed that there is no consensus on the definition of rationality but took issue with the idea that neuroscience could help us define it. The criteria for rationality are socially defined, he argued, and neuroscience could look for consistent ways of acting "rationally" but could not define the thing itself.

Morse separates the free will aspect from the responsibility aspect, arguing that the law is not concerned with free will, but only with responsibility. Indeed, he goes so far as to argue that the debate about mechanistic causation is essentially moot. Law and responsibility are human constructions that are mind dependent, and since we are constrained by our view of ourselves as rational agents, our constructs will reflect these views.[11] In fact, when considering issues of responsibility, Morse argues that it would make more sense to focus on rationality than intentionality, holding that those injuries or deviations of brains that impair rationality are more likely to be viewed as mitigating responsibility than whether the person intended to act a particular way.

Not all participants agreed with Morse's and Gazzaniga's analyses, and some felt that while the idea of free will is unlikely to be discarded, developments in neuroscience might result in substantial changes in society's

---

[10] In his examples regarding the concept of rationality, Morse focuses on the ability of the individual in question to perceive, recall, and understand the facts (this would exclude hallucination or delusion), to be able to know what is relevant, to have a relatively consistent preference ordering, and to know by what rules she is playing.

[11] This is a massive oversimplification of Morse's thesis. Readers are encouraged to read Morse's full argument in his paper in this volume.

concept of responsibility. The challenge raised to the "rules of engagement" view, for example, was framed by one scientist this way: "If even the most disturbed brains can learn some of the simple social rules, are the brains of people who [violate] the social rules [affected with] problems of learning those social rules, and if so, how is the brain responsible?" In response, Gazzaniga made two key points.

First, Gazzaniga argued that the standard for learning rules is fairly high. As he put it, "In other words, for someone not to be penalized seriously, it seems to me you would have to show that they could not learn these rules, and I think that's a very hard criterion to pull off." Second, he noted that a risk arises that "explainable becomes exculpable," and that the possible misuse of neuroscience toward that end should be avoided. In other words, the rule-learning view would allow for the person with advanced dementia, who not only cannot learn rules but who forgets rules previously mastered, to be seen as less culpable than the person who seems to have problems with some rules but who has clearly learned others.

At least one participant felt that in the context of responsibility and free will, it would be helpful to keep separate the ranges of diagnoses and to not talk about "responsibility" as if it were a monolith. In particular, this participant advocated that those "things that look like medical illnesses" be kept separate from "those things that look like traits." Schizophrenia with a severe delusional component, for example, is vastly different in its effect on rule following than, say, a personality disorder. Where one might be inclined toward leniency for the person with schizophrenia, one would not necessarily support leniency for the personality-disordered individual.

Another participant, while not disagreeing with the idea of keeping the types of diagnoses separate, offered a different analysis, observing that the opposite argument was equally supportable. As the classical definition of *personality disorder* is one of a structural, congenital personality defect, it may be argued that the psychopath in general may have less control than many with conditions that look like illness. One could argue that the personality-disordered individual should be granted leniency, perhaps equivalently or more so than schizophrenics, as schizophrenia follows a spectrum from mild to serious delusions.[12]

Many seemed to feel that the greatest impact neuroscience will have on the concepts of free will and responsibility will be felt not in exculpatory ways but in mitigation ("He's not fully responsible, because of his brain") and in perception of risk ("He might try to follow the law, but his brain won't let him"). Some level of this type of influence will almost surely occur, and the concerns will be again twofold: making sure (1) that the neuroscience findings affecting legal concepts are based on sound science, and (2) that such science is used with a cautious eye to the potential societal consequences.

PREDICTION AND SOUND SCIENCE

> *It's not just looking at the technology and saying this technology isn't valid and isn't applicable and isn't reliable. One has to look at that and put [it] into the balance next to "What does the court now rely on in terms of information in making attributions of responsibility?"*
>
> —A lawyer/scientist on the question of the realities of decision making in the law

While "sound science" is the optimal criterion, it is not essential for predictive technologies to be 100 percent accurate (a level of accuracy unlikely ever to be achieved) to be declared of use to the court system. The courts currently use predictive techniques all the time, with very little review of such matters. As one scientist/lawyer participant put it:

> Civil commitment decisions, death penalty decisions . . . decisions are being made every day that involve liberty and life, including substantial loss of liberty, that are based on incredibly weak predictive techniques and

---

[12] The participant recommended two articles for reference: A. Raine, T. Lencz, S. Bihrle et al. 2000. "Reduced prefrontal gray matter volume and reduced autonomic activity in antisocial personality disorder." *Archives of General Psychiatry* 57, 119–127; and A. Raine, D. Phil, J. Stoddard et al. 1998. "Prefrontal glucose deficits in murderers lacking psychosocial deprivation." *Neuropsychiatry, Neuropsychology, and Behavioral Neurolology* 11, 1–7.

virtually never are overturned on the basis that these techniques were insufficient to justify the kinds of decisions that were being made.

Courts use prediction in plea bargaining, sentencing, decisions about levels of probation, and case diversion, among other proceedings. In each of these examples, the courts must weigh future risks, including the likelihood of recidivism, against other societal and pragmatic concerns (like prison overcrowding). Accordingly, to the extent that sound science can better inform those predictions, neuroscience really has something of benefit to offer the court system.

Courts, of course, must make decisions now—they cannot wait for the next round of peer-reviewed research results. This immediacy presses, in some sense, for courts to use any tool that might shed additional light, however dim, on the matter at hand.[13] The risk is that predictive decisions will be based on poor or incomplete science. Additionally, neuroscience-based predictions may be given undue weight as "scientific predictions" when they may still suffer from the typical problems inherent in current risk-prediction models: bias in the construction of normative or sample groups, reliability or validity issues in the prediction itself, and the inability of a predictive measure to reveal details about the particular individual but only, probabilistically, about the group to which the individual belongs.

PREDICTION AND SOCIETAL IMPACT

The use of flawed or incomplete science and the reliance on seemingly scientific predictions based on more than the science is prepared to support are exactly the kinds of concerns that should be foremost in the public mind when contemplating the potential

---

[13] In regards to what can be considered in evidence, the courts are bound by either the federal or the state rules of evidence. But in the preformal stages of criminal cases (prior to official charges being filed), sentencing, case diversion, and other important decisions, the courts have greater latitude. It is here, in nonevidentiary ways, that the greatest risks from the use of poor or incomplete science may arise. See also "Preformal Usage," in Chapter 4.

social impact of predictive technologies or techniques. It is not just in courtrooms that prediction would have an impact. As both Tancredi and Greely point out in their papers, the use of behavior-predictive technologies by schools, employers, health care systems, government investigators, and the like would dwarf usage by the court system. The potential to pigeonhole, to discriminate, and to judge on the basis of test results could result in substantially negative consequences, including the development of a "neuroscientific underclass" denied access to education and other societal benefits on the basis of their neuroscience test results. The parallels, as Greely points out, to the current dialogue on genetics are instructive, and those discussions may well illuminate some of the promises and pitfalls that could accompany a greater understanding of the brain.

At the meeting, Greely postulated four overarching points about prediction that are instructive: (1) if the correlation is strong enough, it does not matter what the causative mechanism turns out to be[14]; (2) people tend to jump far too quickly to finding a "strong" correlation between things[15]; (3) even when we do have strong correlations, they will not be 100 percent correlated, which is something people generally have a difficult time accepting[16]; and (4) people often get excited about high-tech tests, but when cheaper and better ways to test are available, the high-tech tests may not be used. (See "Paper and Pencil," page 16.) These four points, derived from Greely's experience with genetics, serve as useful touchstones when considering the potential impact of neuroscience-based predictions.

---

[14] Or, in an alternative phrasing, it does not matter which way the "causal arrow" points.

[15] Greely's experience with genetics is instructive here. As he said during his presentation at the meeting (referring to reported "strong" correlations between genetic variations and schizophrenia that turned out to be nonreplicable), "So things often turn out to be wrong, despite the fact that they were done completely nonnegligently and nonfraudulently. It takes a while to get the science right."

[16] Greely's example here also comes from genetics, where "we see people acting as if 60 percent probabilities are 100 percent probabilities, and 10 percent probabilities are zero probabilities."

Paper and Pencil

§

For many of the states or traits likely to be of interest to the legal system, there are already existing tests that do not involve neuroscience imaging or testing, but instead are simple "paper and pencil" instruments. Many conference participants raised the point that the availability of these relatively inexpensive and easily portable tests may serve as a natural check on the rate at which neuroscience moves into the courtroom.

The most common example given was Hare's Psychopathy Checklist (PCL), a test for diagnosing psychopathy. The PCL is widely used, is well known by clinicians and researchers, and has been used extensively in research settings, so there is a large body of knowledge about the instrument itself.[†] As one scientist observed, "To the extent that scans are used now to try to uncover neurocorrelates of psychopathy, they're validated against the PCL." Until neuroscience scanners become more portable, until there are more data to draw upon, and until the per-use cost comes down, the use of neuroscience tests may not necessarily pose a serious challenge to tests like the PCL.

Two Counterpoints Worth Considering

First, as one scientist at the meeting put it: "Are [neuroscience tests] just a different and new and very expensive and flashy way of getting certain kinds of information or do they really offer something qualitatively new and different and better than previous

---

[†] While the use of the Hare PCL is common in a variety of settings, it is not without controversy. The definition of *psychopathy* itself is quite controversial, no less than the construction of any scale to measure it. The PCL is used here simply as an example of the type of paper and pencil tests discussed, and it is one that is commonly available. See Erica Beecher-Monas and Edgar Garcia-Rill, *Danger at the Edge of Chaos: Predicting Violent Behavior in a Post-Daubert World*, 24 Cardozo Law Review 1845 (2003). Also R. D. Hare, *The Hare PCL-R: Some issues concerning its use and misuse*, 3 Legal and Criminological Psychology 99–119 (1998).

methods?" If neuroscience tests are qualitatively different, and offer better information, they will likely become increasingly attractive to the legal system.

Second, there were concerns that neuroscience tests may be seen as more "substantial" evidence by juries, even when the state of the science would not support such a claim. Although there may be existing "paper and pencil" tests, structured interviews, or behavioral assessment tests to assess the construct of interest, the neuroscience scans and tests will be assumed to be more objective, scientific, and accurate than they perhaps are. The apprehension is that the high-tech biomedical appearance of the neuroscience tests will carry undue weight with juries, especially jurors whose expectations about the use of forensic science have been conditioned by television courtroom dramas.

Predicting Violence

Though a host of possible predictions might be desirable (for example, predictions about the tendency to be honest, the willingness to follow authority, and so forth), the potential for predicting violence is of particular interest and significance.

The prediction of violence has already been the subject of some neuroscience research, and it will probably continue to interest science as well as the legal system. All the previous discussion on behavior prediction is directly relevant to the prediction of violence: it is a predictive measure likely both to have tremendous utility and to carry great risk of misuse, and it is likely to cut both ways in criminal law—in mitigation and in marking someone as being predisposed to violence. While violent behavior will probably never be predicted with complete certainty, the likelihood that techniques will be developed to distinguish those more likely (or even very likely) to react with violence seems great enough that the social impact of those techniques should be considered for future research and public discussion.

An additional concern is preemptive uses of prediction of violent behavior (or of proneness to violence). Generally, in the legal system, we punish people based on behavior, not on thoughts or "tendencies."[17] The idea of imposing treatment, or even making decisions regarding employment, based on some test results, and in the absence of prior violent behavior, is deeply disconcerting.

Of course, not all the possible ways in which predictions of "violence proneness" could be used are negative. For example, in screening people whose jobs require them to confront violence[18] and, in some circumstances, to respond with violence, such tests may be extremely useful. This might be thought of as identifying "violence eligible" individuals.

Competency and Cognition

Though it may not be immediately apparent, determining whether someone has the capacity to act in a legally binding or efficacious way (for example, to sign a binding contract, or to make decisions regarding her medical treatment) shares some similarities with the prediction of behavior. Both assessments can influence how we think about a person's legal responsibility for her behavior. With the prediction of behavior, as discussed above, we may be willing on rare occasions to say that a person cannot learn the rules of society, or cannot follow those rules, and so may be less responsible for her actions than someone able to learn and follow those rules. Similarly, regarding the capacity to act, we may look for diminished ability to do what the law expects or requires, and then either release a person from a legal responsibility or prevent a person from exercising an option he no longer has the capacity to perform.[19]

---

[17] One notable exception is cases in which people have already committed violence, and the question is the likelihood of committing further acts of violence. We do this constantly in the criminal justice system. However, we do not do it prior to an initial act of violence, and we generally reserve such predictions for the most violent offenders.

[18] For example, members of the armed forces and law enforcement officers.

[19] One example involves the formation of a contract. If the signatory to a contract could not understand what he was doing, due to advanced dementia, then he may be held not liable for his duties under the contract, even though he freely signed it. Similarly, a person with dementia may be prevented from contracting by having a guardian appointed.

In general, the law assumes that adults can act in a way that has legal effect—for example, they can bind themselves to contracts, manage their affairs, or consent to medical procedures. Historically, as Tancredi discusses in his paper, a largely phenomenological psychiatric examination has been used to determine whether someone lacks the mental capacity to exercise his legal rights and be held to his legal duties. Tancredi speculates that advances in understanding memory processes, neural circuitries, and the relationship of genetics to neurological function may help in the development of increasingly sensitive and accurate methods of evaluating competencies, particularly when paired with advances in neuroimaging. In addition, medications developed for the treatment of memory disorders may also help to address competency issues from the perspective of clinical treatment. For example, medications that facilitate the ability to remember or process information might "restore" capacities the lack or diminishment of which might result in a person's being adjudged "incompetent" in a specific context.

Looking Past Words: Neuroscientific Lie Detection

The development of accurate neuroscience-based lie detection would be one area of brain monitoring or imaging with immediate value in the law. Tancredi goes through some of the techniques and studies currently being developed, none of which are ready for widespread use. Among the technologies being explored are the use of near-infrared brain scans that focus on activity in the prefrontal cortex; thermal imaging, which looks for increased heat emission around the eyes; computer programs that analyze faces for subtle, nearly imperceptible changes; MRI that focuses on increased activity in the anterior cingulate cortex and superior frontal gyrus; and the use of the electroencephalograph (EEG) to look for a particular brain wave pattern, the P300 wave. However, one participant noted that the aforementioned techniques are not based on a clear neuroscientific understanding of the phenomenon of lying. The lack of any underlying cohesive theoretical framework means that the current work relies solely on correlation, essentially "shooting in the dark."

Of all these techniques, the only one that has been considered for admission into evidence is the EEG/P300 technique, which has been marketed as Brain Fingerprinting®. The P300 wave occurs when the subject recognizes information or a familiar stimulus. The examiner, in essence, reveals detailed and specific information about the crime to the subject, and observes whether the P300 wave occurs, suggesting familiarity with information known only to the person who committed the crime (and, obviously, the investigators and the examiner). While the Brain Fingerprinting evidence was not dispositive—that is, it did not determine the outcome of the case—it was admitted into evidence and considered by the judge in a state court case in Iowa.[20] As this is apparently the first case in which this evidence has been admitted, it remains to be seen if it will be judged admissible in other Iowa courts, or in other state or federal courts. It seems reasonable, however, to suppose that eventually such technologies will be developed and relied on.[21] If accurate and reliable, such tests could find their way into the courtroom, but how that might occur and with what effect is still unclear.

The most significant hurdle facing accurate lie detection is what could be termed "the problem of memory." While it seems likely that techniques might be developed to detect when someone is intentionally

---

[20] The trial in which the Brain Fingerprinting technique was used was an appeal from a murder conviction. The defendant, who had been convicted of murder some 25 years prior, was arguing for a new trial on a number of grounds, including that the Brain Fingerprinting showed that he did not recognize key aspects of the crime. While the Iowa District Court upheld the defendant/appellant's sentence, the Iowa Supreme Court subsequently reversed the lower court's ruling, ordering the conviction vacated and a new trial for the defendant. See *Harrington v. State*, No. 122/01–0653 (Iowa Sup. Ct., 2003). While the ruling of the Iowa Supreme Court did not turn on the Brain Fingerprinting, it was specifically mentioned in the court's discussion of newly discovered evidence cited in the appeal. Subsequent to the reversal and remand for new trial, the prosecution announced it was dismissing the case, stating that the "admissible evidence which is left after 26 years is not sufficient to sustain a conviction against Mr. Harrington." Accordingly, after serving 25 years in prison, the defendant was freed. See Mark Siebert, "Free Man," *Des Moines Register*, October 25, 2003. This turn of events will likely bring greater attention to the use of the Brain Fingerprinting technique in future trials.

[21] The notoriously inaccurate polygraph is currently relied on in a variety of contexts, although it is not admitted in courts as evidence. Some of the uses of the polygraph have what could be considered "legal effects," as they can affect people's rights and opportunities—such as the use of the polygraph by employers to screen employees.

lying, several scientists expressed doubt that one could easily detect when someone is merely mistaken—that is, when someone is subjectively telling the truth but is factually wrong. While this may be a hurdle that will one day be overcome, scientists viewed it as a substantial problem common to many of the scientific approaches to lie detection.

Even if the science becomes able to provide 100 percent accurate lie detection, numerous legal questions will still need to be addressed. While Greely covers these issues in more detail in his paper, two legal issues are undeniably important: the role of the jury and compulsory testing of witnesses for veracity.

The concern about using such tests, if they become available, is that the evaluation of witnesses, and the credibility of and weight given to their testimony, are matters for the "finder of fact," the body charged with determining the facts in the matter before the court. In jury trials, this is the jury; in bench trials, it is the judge. In allowing scientific testimony regarding truthfulness into evidence, the court may well be invading the purview of the jury. Even if admission of such evidence is not held to invade the purview of the jury, there will still be concerns about whether such evidence may have undue influence. Members of the jury may weigh "scientific evidence" more heavily than their opinion as formed by their own senses, and may do so specifically on the matter of truth. Typically, the matter of credibility of witnesses has been held to be a core function of the finder of fact, and a determination that should rest on the evidence of the finder's own senses, and not a matter of expert testimony on truthfulness.[22]

The other concern, regarding the compelling of witnesses to be tested for truthfulness, involves several factors. An obvious one is the testing of defendants or potential defendants. Aside from concern about the Fifth Amendment's safeguard against compelling testimony against one's self, should a judge or jury be allowed to consider a defendant's refusal to take such a test? Greely points out in his paper that a refusal to be tested may destroy *any* witness's credibility, particularly when other witnesses

---

[22] See *United States v. Scheffer*, 523 U.S. 303 (1998); see also Greely's discussion of this case in his paper.

telling a different story have agreed to and passed the test. As polygraph tests are rarely admissible, the challenges facing the accurate testing of defendants have yet to be examined, but with accurate neuroscience-based lie detection techniques, they might come to the fore.

Similar questions apply to nondefendant witnesses in trials both civil and criminal. Could a party subpoena a witness and demand a lie detection test? While safety concerns would be an obvious issue,[23] so would issues of privacy—what else might be learned, what else might be asked? While what can be asked in criminal trials can be quite limited, the scope of, say, civil depositions can be quite broad. Could a person be compelled to answer a subpoena while being monitored for veracity? Does one have the right not to be so tested? Clearly, issues abound. In addition, uses of lie detection technologies in employment contexts and in the debriefing of intelligence agents, have potential legal impacts.

DETECTING BIAS

Closely related to the issue of lie detection is the detection of bias. Some early research has shown that some brain activity can be detected when people who have bias toward certain groups are shown pictures of members of those groups. While such research is in the early stages, again the potential uses seem broad in scope: jury selection, discrimination cases, and employee screening, just to name a few. The difficulty arises in determining exactly what the person being tested is reacting to, and why. While the brain activity may be correlated highly with bias, it may correlate with other beliefs or states as well.

---

[23] Several participants noted that many of the technologies discussed (though not all) are sufficiently new or invasive that concerns about the safety of the procedures cannot be casually dismissed. (For example, when imaging the brain using positron emisson tomography—or PET—scans, radioactive contrast material is injected into the body. While the risks of such procedures are generally considered to be low, there is some risk nonetheless.) It is one thing to undergo a test or procedure to diagnose a potential health threat; it is another to undergo that test or procedure because of entanglement in a lawsuit. The potential benefits in the former situation may outweigh even substantial risks, whereas in the latter any nontrivial risk may be judged too high.

The concept of detecting bias raises an issue that also permeates the consideration of prediction. To what extent do we, as a society, wish to judge people based on what we perceive they are thinking rather than what they say or do? This is not a trivial matter, and a comprehensive discussion of its implications is beyond the scope of this report. However, it is worth noting briefly that we typically reward, punish, or hold people responsible for their actions, not for their thoughts (or potential actions). In spite of our general "actions, not thoughts" philosophy, *once we have decided to punish or hold someone responsible,* we routinely place restrictions on the person based on "predictive" assessments (that is, we restrict people based on what they *might* do).

Because the matter of assessing bias generally will not involve the issue of punishment or criminal behavior (unlike with the prediction of violence), the decision-making rules might be somewhat different. For example, the tension between the principles of "actions, not thoughts" and "predictive assessments" may pull toward the former when a decision is being made whether to hire someone who might have a bias (that is, where the risk is primarily one of having to deal with biased behavior, if it were to emerge), and toward the latter when the issue is whether to seat someone with a bias on a jury. Where the risk to the defendant (or any particular individual) may be substantial, we may feel more comfortable erring on the side of caution.

Brain Death

The final topic on monitoring is the determination of brain death. The question is whether the definition and determination of brain death might be better informed or substantially changed by neuroscientific developments. Tancredi goes into this issue at length in his paper, tracing the history of the development of brain death standards. The current standards are primarily focused on brain-stem death—the areas of the brain that deal with the automatic processes of the body, like respiration, blood pressure, and heartbeat. Tancredi notes that improvements in understanding and monitoring brain function may well influence definitions of brain death by focusing attention on higher cerebral function. As he phrased it, "It may be more intelligent to have brain death be when higher cerebral functioning is no longer occurring because we no longer have the person." The underlying idea is that neuroimaging would show when higher cerebral function has permanently ceased, and thus indicate when speech, cognition, learning, consciousness, and other defining human characteristics are irretrievably lost.

One constraint on the impact of new neuroscience in this area is the extent to which substantial work has already gone into defining brain death, with input from many different groups and with a substantial moral and religious literature to draw on in conceptualizing and defining the state of being "brain dead." The comments of one participant reporting on a small-group discussion at the meeting mirror a consensus view of the participants:

> As we discussed the brain death issue, we realized that the whole current system for identifying death represents an artfully drawn balance between a lot of legal, social, and religious issues, and we're very apprehensive about [altering] that. So while there might be opportunities certainly to explore and do research in this area, it certainly is an area that deserves a lot of caution, and we did not feel that this should be the most prominent area of research in the near future.

3

Modifying the Brain

As we learn more and more about the brain . . . it will give us more and more control and more and more ways in which we can manipulate the brain, and I think that is going to be one of the most serious questions. When, under what conditions, and who decides what manipulations of somebody's brain are permissible or not?

—Neuroscientist participant

WHILE ONE MIGHT reasonably expect that the monitoring aspects of brain science would precede the modifying technologies, it turns out that some modifications are already available and may soon pose serious legal questions. On the near horizon are what may generally be termed enhancements, but a modification currently available—the pharmacological treatment of addiction—already raises a multitude of issues.

Enhancements

The opportunities are in front of us for the first time perhaps in history to really enhance man's capacities in the cognitive areas to levels never before attained, providing possibilities for higher IQs, greater memory power, acute

sensitivity to the nuances of human communication, and
heightened awareness of one's moral responsibility.
     —Laurence Tancredi, in his introductory remarks

While we may think of enhancements as only major changes to the brain—cell transplants, chip insertion, and the like—in reality, many of these changes are so far off as to make it impractical to address them. However, some enhancements of a pharmaceutical nature are immediately available, and they can serve as a template for contemplating some of the larger issues of enhancement.

From Provigil, which allows people to go days without sleep, to the latest drugs that enhance the retrieval of memory in people suffering from dementia, pharmacology offers many opportunities to consider questions of enhancement. An area with significant potential for controversy is the off-label use of drugs such as Ritalin to improve attention and performance in scholastic testing.[24]

Many of the legal issues raised by considering enhancement are less a challenge to courts and laws than to the larger policy questions: distributive justice, disadvantaging effects, and the potential for creating an unenhanced underclass. The general concern may be that those with privilege will seek enhancement to develop a competitive advantage over less privileged individuals.

The example of the Scholastic Aptitude Tests (SATs) illustrates these issues clearly. Should we test students for Ritalin use immediately after the SATs and, if they test positive, void their scores? Have the students using Ritalin taken an unfair advantage? Or is performance enhancement with a drug comparable to SAT prep courses, which may provide students advantages over those without access to such courses? The difference between the two, it may be noted, is that the prep course requires effort. In addition, the use of drugs (even very safe ones) may entail health risks.

---

[24] Much work is currently being done to develop drugs to enhance the memory of elderly persons with memory impairments. The market for medical treatment will likely drive the development of drugs that then "cross over" to enhancement uses.

Reflecting on Ritalin and the SATs highlights a strong aspect of the American legal system: the emphasis on individual rights will make it difficult to ban or restrict enhancement technologies simply because they may disadvantage those without them. With that background, consider the opposite: When can enhancement be ordered?

NOT SICK, BUT BETTER

Mandated enhancement becomes a question in the following context: When could you ask or compel someone to take a selective serotonin reuptake inhibitor, a class of drugs commonly prescribed for depression (among other disorders), in order to make that individual less angry, less impulsive, and less irritable, but in the absence of a psychiatric diagnosis?[25] Could we ask or compel prisoners to do it? Could it be a condition of probation? While this could very well prove to be legally permissible, say, as a condition of probation (though not permissible as part of imprisonment), it still raises substantial ethical concerns, as well as some potential constitutional issues about the "integrity of the person" and the extent to which the state could interfere with the functioning of an individual's mind.

Typically, the extent to which the government is allowed to force psychiatric medication on a person is extremely limited. In most circumstances, the person must be a threat to himself or others before the state (typically through a brief court hearing) can compel him to take medication (in almost all cases we are talking about psychiatric medication, typically antipsychotic drugs). This is true for prisoners as

---

[25] For the purposes of this report, we are using a simple metric for separating treatment and enhancement. Where there is a diagnosis, the use of medication is treatment; in the absence of a diagnosis, medication is being used to enhance some aspect of the person. This is a unsophisticated definition, but workable for the task at hand. For a more nuanced treatment of enhancement issues, see E. Parens, M. J. Hanson, and D. Callahan, eds., *Enhancing Human Traits: Ethical and Social Implications*, Washington, D.C.: Georgetown University Press, 1998.

well, though arguably the high-risk environment of the prison means that it is fairly easy to find the person is at risk of harm.[26]

In the case of criminal trials, the court may order antipsychotic drugs to be administered to a mentally ill defendant, without that person's consent, in order to render the defendant competent to stand trial on serious criminal charges. Before the court can do so, it must be satisfied on several points: the treatment is medically appropriate, the treatment is substantially unlikely to have side effects that may undermine the trial's fairness, no less intrusive alternatives that might be effective are available, and the treatment is necessary to further important governmental trial-related interests.[27]

Unlike the individuals in the situations described above, the person who is the target of enhancement efforts does not, by definition, suffer from an illness but falls within the normal range of behavior. The real ethical and legal concerns of mandatory enhancement may involve not court-ordered medication but "soft" coercion by the state to "voluntarily" take the medication (by making it a condition for early release from prison, for example).

THE FUTURE OF ENHANCEMENT

Tancredi's paper ranges far afield in posing theoretically possible neuroscience breakthroughs beyond what we might expect to occur in a reasonable time frame. These thought experiments include downloading memory into a computer, inserting knowledge (or, viewed another way, uploading "memories") into the brain, brain transplantation, and the reversal of apoptosis (natural cell death) in the brain. Such dramatic

---

[26] See *Vitek v. Jones*, 445 U.S. 480 (1980). The court ruled that for the state to be able to treat a prisoner against his will, the inmate must (1) suffer from a "mental disorder" and (2) be "gravely disabled" or pose a "likelihood of serious harm" to himself or others.

[27] The governmental ability to medicate defendants without their consent has been set out by the United States Supreme Court in three cases: *Washington v. Harper*, 494 U.S. 210 (1990); *Riggins v. Nevada*, 504 U.S. 127 (1992); and *Sell v. United States*, 123 S. Ct. 1385 (2003). See also *Cruzan v. Director, Missouri Department of Health*, 497 U.S. 282 (1989), on the right to refuse medical care generally.

examples help to frame thinking about enhancements that will be acceptable to society, and those we might find objectionable, to the point of enacting laws or declaring their use to provide unfair advantage.

While the examples listed in the paragraph above may sound so dramatic as perhaps to be unrealistic, they should not be dismissed out of hand. If there is a single lesson to be learned about the past century of scientific and technological discovery, it may well be that the unimaginable rapidly becomes the commonplace. For example, one enhancement technology showing promise in its early stages is transcranial magnetic stimulation (TMS). TMS uses pulsed magnetic fields to inhibit or enhance the functioning of a selected area or areas of the brain. This is done without invading the brain, through a helmet that generates and focuses the magnetic fields for the desired use. Researchers can, for example, create a temporary "stroke patient" by using the TMS helmet to inhibit the area of interest. Similarly, they can boost functioning of an area for a temporary form of enhancement.[28] Smaller, more portable versions of the TMS helmet are already under development for a variety of purposes, ranging from the treatment of depression to keeping sleep-deprived soldiers awake and alert.[29]

Treating Addiction

One type of enhancement with immediate importance and pertinence is modification of the brain to treat addiction, particularly addiction to opiates. Treating addiction is viewed by many as a long-term, if not lifelong, process. The relapse rate is high, and the legal penalties for illegal opiate use are substantial.

---

[28] A particularly provocative account of the use of TMS for enhancement purposes appeared in the popular press, recounting a reporter's personal experience in having his ability to draw temporarily enhanced via TMS. While the reporter's account is obviously not a peer-reviewed study, and therefore does not provide the scientific community with the level of precision desirable, it does offer an excellent example of how the general public is likely to become informed about and view these technologies as they are developed. The title of the article really says it all. See Lawrence Osborne, "Savant for a Day," *New York Times Magazine*, June 22, 2003, pp. 38 *et seq.*

[29] Ibid.

Neuroscientists working in this area have demonstrated that the brains of addicts are different from those of nonaddicts and have found evidence of a genetic predisposition toward addiction.[30] In considering responsibility and addiction, the idea of a different brain state is intriguing. As one scientist observed, when it comes to addiction, other factors may be so salient as to upend rationality. A striking example is the radical reordering of priorities that often occurs in addicted mothers of newborns who abandon their babies in the hospital to go get drugs, their addiction completely overriding their maternal instinct. As one scientist commented:

> So then the question is, is this free will? . . . Can you really use the same term of some sort of free choice involved when a mother actually deserts her baby? It just seems to me that it strains our definition of free will, and I would have to say that there is some sort of compulsion that develops with the illness of addiction that sort of overcomes what we normally think of as free will.

As it turns out, highly effective pharmaceutical treatments for opiate addiction, with few or no side effects, are currently available and yet not widely used. Here is a clear example where neuroscience could directly influence and affect law, but it has not. One drug, naltrexone, serves to block the pleasurable or rewarding effect of the opiates.[31] By blocking the receptors to which the drugs bind, the medication makes relapse

---

[30] One neuroscientist participant observed, "There are many different kinds of addiction, but . . . it's very much influenced by heredity and even smoking actually has very powerful hereditary influences, and even a powerful drug like cocaine—only 16 percent of those who try it actually become addicted. So most people are able to resist it and among those who become addicted, the free will clearly is in the first dose. There are some drugs that almost everybody tries, like alcohol. Ninety to ninety-five percent of people try it, but only about 9 or 10 percent become addicted to it . . . "

[31] Another possibility is emerging in medical trials involving vaccines that block the rewarding aspects of drug use and thus may be used to treat drug addiction. For example, one vaccine, TA-CD, reduces the euphoric effects of cocaine. See "Cocaine Trials Progress," BBC News, April 2, 2002, at http://news.bbc.co.uk/1/hi/health/1906823.stm. Similar vaccines for treating nicotine addiction are also undergoing test trials.

impossible as long as the individual continues to take it. The clinical problem is getting the person to comply with the medication schedule. Compliance could be ordered as a condition of probation or parole. Such mandated adherence to medication would be facilitated by a preparation of naltrexone that would act over a prolonged period, and one is currently being considered for approval by the FDA. As long as former opiate addicts were required to take a monthly injection, they could not relapse into opiate addiction. Successful drug treatment not only reduces the health risks associated with drug use but also eliminates the legal risk of incarceration for possession of drugs or drug paraphernalia. Drug addicts could, in theory, be diverted to a mandatory treatment program at a much lower cost than incarceration.

The naltrexone example raises the question of how such a discovery can result in changes in law and policy. At least one participant foresaw the depth of change involved in incorporating such a finding into our current society:

> Well, right now, rehabilitative models don't play much of a role in punishment practices, but if we all of a sudden started to get [addiction treatment] that really worked, they very well might . . . [of course] this is always going to be subject to further understandings of our morals, our politics, and our science . . .

The issue of treatment for addiction perhaps serves as the best example of the need for efforts whereby lawyers and scientists seek to inform each other's work. While it is not obvious how this particular issue might change the law, clearly a continuing dialogue should be maintained,[32] and a greater effort should be made to facilitate education and interaction between the neuroscience and legal fields. Real, and possibly immediate, benefits may be gained from such efforts.

---

[32] One group of participants suggested that the scientific and legal communities go further in calling for the effective treatment of addiction: "There seems to be a lot of potential there that . . . we felt did not receive proper recognition. This [currently available medications and techniques to help deal with the problems of addiction] is something that would certainly be a benefit to the criminal justice system, if it is provided as an opportunity for treatment . . . ."

Crosscutting Legal Issues

A s noted at the beginning of Part I, a number of legal issues cut across the monitoring and modifying distinction this report has used to consider neuroscience developments. Five such crosscutting issues arise in a variety of contexts: discrimination, privacy and confidentiality, preformal uses of neuroscience, litigation-related issues, and intellectual property. Just as the science is still developing, the current state of the law regarding neuroscience specifically is even more nascent. Still, to the extent that antidiscrimination statutes may restrict how neuroscience techniques are used, or that intellectual property concerns may have a limiting effect on some research, some brief consideration of the potential impact of the law on neuroscience is also appropriate.

Discrimination

> *Tests are important in our society. There's an actuarial mind-set that occurs in our society for prediction of risks, and a framework for assessing costs. So we have to assume [tests] are going to happen. They are going to happen in the academic world. They are going to happen in the legal system . . .*
>
> —Laurence Tancredi, in his comments on the risks of discrimination

32

The issue of discrimination is a concern with both monitoring and modifying technologies. The use of monitoring technologies, particularly in predictive applications, could lead to the pigeonholing of children, the denial of opportunities, and other forms of discrimination or "neurological prejudice," even as limited access to modification technologies could produce a growing divide between those with access to enhancements and those without, creating a "neurological underclass."

There are similarities between apprehensions about discrimination based on neuroscientific tests and procedures and unease about discrimination based on genetic tests, procedures, and information. In both cases, the concern is that people will be disadvantaged based on their biological makeup—either of brain or DNA—rather than on their own demonstrated abilities and accomplishments. Interestingly, while genetic discrimination has engendered a fair amount of activity—including congressional hearings, proposed federal legislation, some enacted state legislation, and numerous meetings by various august bodies—very little litigation has arisen to date. Possibly, cases of neuroscientific discrimination could be similarly slow to develop, with the legal and scientific communities leading the charge to examine the issue before such cases arise. Work being done currently on genetic discrimination may serve as a good model and guide for future efforts to prevent neuroscientific discrimination.[33]

One specific issue arising directly from genetics may be even a greater risk in the neuroscience context—the idea of "exceptionalism." In genetics, a concern has been voiced that passing laws and special rules for cases of genetic discrimination (rather than treating such matters under current antidiscrimination schemes such as the Americans with Disabilities Act of 1990 [Pub. L. 101-336]) will result in a perception

---

[33] One participant noted that a particularly unusual aspect of the genetic discrimination issue was the strong interest from the federal executive and legislative branches, as well as from state legislatures, even though there has been almost no litigation. Typically, court cases precede legislative efforts in matters like this, but with concerns about genetic discrimination, there has been an unusual level of interest and activity. This participant suggested that it was a real possibility that the issue of neuroscience-based discrimination may engender a similar response, one for which the participants may want to prepare.

by the public that genetic factors are more important and determinative of our well-being and behavior than, in fact, they are. Singling out genetic information for special protection seems to indicate that an exceptionally powerful amount or type of knowledge is there—hence, genetic exceptionalism. Similarly, special legal protections against discrimination based on neuroscientific information could lead to a form of "neuroscience exceptionalism"—the idea that information about our brains is more determinative than it is in fact. So while discrimination based on neuroscientific knowledge is a significant risk that the neuroscience and legal communities should work to minimize, it is not clear that a new legal structure specific to neuroscience should be developed, in light of the risk of unintentionally promoting a unwarranted public perception of "neuroscience exceptionalism." An alternative path may be to strengthen existing antidiscriminatory statutes and schemes to include discrimination based on neuroscientific information.[34]

Greely provided an interesting gloss on a similar risk—that of neuroscientific essentialism. Again paralleling a similar concept in genetics, neuroscientific essentialism is the idea that one's essence is one's mind/brain. However, the essentialism argument may not be as erroneous in neuroscience as it is in genetic science. As Greely noted, "It seems to be quite possible that I am my mind, or I am my brain, in a way that I'm quite clear I am not my genes. My genes are not me. My mind, my brain, well, maybe that is me." If most people were to agree with the latter notion, neuroscientific essentialism could drive discriminatory behavior (or concerns about preventing such behavior) in a way that neuroscientific exceptionalism does not. That is, if many people feel their minds/brains truly are the essence of "who" they are, there may be less public resistance to using neuroscientific information to evaluate people than there would be to using genetic information (which people see as not reflective of "who" they are).

A key way to reduce the risk of discrimination based on neuroscientific information, of course, is to limit or restrict access to that information,

---

[34] This idea is echoed in the brief discussion of confidentiality and privacy issues, below.

making it available only to the appropriate individuals for uses that society deems acceptable. Accordingly, this presents issues of privacy and confidentiality.

Privacy/Confidentiality

The issues of privacy (keeping information that one does not want known from being discovered by others) and confidentiality (keeping information that needs to be disseminated from going to unintended recipients) regarding neuroscience information have strong parallels with the concern about genetic information. Much of this information, since it would be gathered in a health care context, would most likely be protected, like other health care information, under the federal Health Insurance Portability and Accountability Act (HIPAA), as well as by state laws regulating the confidentiality of medical information. The remaining information would probably be gathered in a research context and would be subject to the confidentiality protections covering research subjects, including the anonymizing of data. So, while the risk of possible breaches of privacy and confidentiality is an important concern, some good protective measures are already in place, and vigilant enforcement of those protections may preclude most problematic disclosures.

However, one issue the current standards and practices may not be prepared to address is the scope of information gathered. Compared with a blood test for the presence of a specific antibody, the amount of information gathered in a single imaging procedure is considerably broader in scope. Therefore, when testing for one particular characteristic or marker, substantial additional (or collateral) information is gathered. This collateral information may be sufficient to identify other characteristics, markers, or conditions that the person being tested would like to keep private, or that could even become the basis for discrimination.

The difficulties of dealing with collateral information are already being felt by the research community. With respect to the compilation of large databases of MRI and fMRI data for brain research purposes, concerns have been voiced that the additional (in this case, nonbrain) data might allow for facial reconstruction of the subjects, leading to their identification and thereby breaching confidentiality. The immediate solution has been to scramble the facial data,[35] though research data are lost in this procedure. How to deal with this aspect of neuroscience testing appropriately has yet to be fully addressed.

Additional difficulties that are likely to arise with regard to collateral information are issues surrounding both informed consent and court-ordered testing. For informed consent, the issue is similar to that in gathering DNA data—potential future uses and discoveries, the possibility of detecting something the subject may or may not want to know—the scope of what, exactly, the subject is consenting to. For researchers this can be a serious impediment to developing appropriate consent procedures and content. For example, what will researchers do if a head scan (say, in a study of variations in the activation of Broca's area during different speech tasks) reveals plaques on the brain suggestive of multiple sclerosis or symptoms indicative of early-stage Alzheimer's? Are subjects consenting to waive being informed, as those diseases are not part of the study? While many of these research issues are ethical and not necessarily legal, they have potential for future legal components or actions[36] and thus must be considered by both communities. Similarly, in the context of court-ordered testing, concerns about what exactly is being learned about the individual would be very important, particularly when the individual is a criminal defendant. While the potential reasons that a court could order testing would vary (competency or mental functioning, for example), the

---

[35] "Open Your Mind," *Economist*, May 23, 2002, reports that the managers of the fMRI Data Center at Dartmouth College are scrambling facial information in the scan data they are compiling, making the data "useless for some forms of analysis." For more on the database, see Eliot Marshall, "A Ruckus over Releasing Images of the Human Brain," *Science*, 289 (September 1, 2000): pp. 1458–1459.

[36] One conceivable legal action would be litigation stemming from a decision by researchers not to reveal to the subjects aspects of the collateral information gathered about them, when such aspects subsequently appear to have negative health consequences.

concern would be the same—that collateral information collected might be misused.

As one neuroscientist noted:

> A difference between some of these scanning techniques and the more traditional techniques for getting personal trait and state information is that you can obtain information in a scanner that the subject doesn't know you're obtaining. So, for example, if you have somebody sit down and fill out a personality questionnaire, it's pretty obviously a personality questionnaire, and they may or may not want to cooperate and give you their clearest picture of what their personality is like.
>
> On the other hand, [there's nothing about] the scanning methods that have been used to uncover markers for personality that makes it obvious that it's a personality test. They certainly require subjects to comply and cooperate. You've got to lie still in the scanner and you've got to keep your eyes open, look at the pictures, and so forth, but you could certainly be led to believe that something else is being tested.

Finally, Greely suggested an intriguing possibility that hinges on the development of unintrusive detection technologies that can be built into the security arch in airports, courtrooms, and other secured environments. The use of concealed neuroscience technology may raise a host of new privacy concerns depending on the contexts in which it is used.[37] Since unintrusive detectors are unlikely to be developed in the immediate future, there is no pressing need for specific approaches for dealing with such developments, but, as Greely suggested, such technologies would probably need to be regulated to prevent abuse by nongovernmental actors.

---

[37] For example, one could envision such technology being used not only in airports to predict propensity to commit violence in the near future, but also in prisons, schools, psychiatric facilities, and other settings where one's presence or attendance is not entirely voluntary.

Preformal Usage

> My greatest fear, both in the behavioral genetics area
> and with [neuroscience], is that there are no [rules
> of evidence] that control the use of these kinds of
> technologies in the preformal stages of criminal processes.
> When it gets to the formality of sentencing, the cry will
> come up, but the ability of judges and prosecutors to make
> decisions about whether they're going to initiate charges,
> [whether] they're going to accept diversion [from criminal
> prosecution] for people, et cetera—using [neuroscience
> tests] that haven't been validated—is a serious risk that
> the technology poses.
>
> —Neuroscientist participant

While many of the uses of neuroscience addressed in this report occur
in "formal" contexts (in a lawsuit, competency proceeding, or criminal
trial, for example), a cause for concern could be how neuroscience might
be used by the legal system in "preformal" ways—particularly prior to
bringing criminal charges, but in other situations as well. For example,
defense counsel could bring test results to prosecutors as part of a
precharging dialogue, seeking dismissal, reduction of charges, or some
other outcome. Such uses would essentially be unreviewable, and possibly
nonpublic. While the exact nature of these preformal uses is unclear, it
seems prudent that both lawyers and neuroscientists be aware of how
they might be addressed in ways that are both legally and scientifically
appropriate. The scientific and legal communities may even wish to
begin an explicit dialogue on standards, knowledge, and scientifically
appropriate uses.

Litigation-Related Issues

While several neuroscience developments reviewed so far have
aspects that relate to litigation (and are discussed in appropriate
places in this report), some overarching aspects apply generally to

the use of neuroscience in all litigation situations. Greely distilled four key questions:

- First, will the scientific information or test meet the admissibility requirements of *Daubert, Frye*, or other (state) standards? (See *Daubert*, page 40.)
- Second, even if admissible under *Daubert*, are there other reasons that compel the courts to preclude the information?[38]
- Third, should the justice system allow into evidence someone's testimony of his willingness or unwillingness to take a neuroscience test of some sort?
- Fourth, should society ever compel a witness or other person in the litigation process to be tested, and if so, under what circumstances?

These four questions form an admirable précis of the central issues regarding the use of new neuroscience techniques and knowledge in the courtroom. Beyond the admissibility concerns addressed in the first two questions (and expanded upon in the sidebar), the third and fourth questions really must address the as-yet unknown quantities of these new techniques. The questions have been asked before, in reference to the unreliable polygraph. However, if reliable and accurate neuroscience tests are developed that serve legal needs, these are questions that will have to be answered—with the concerns about collateral information, safety, and other risks discussed in this report at the forefront of the debate.

---

[38] The example suggested by one participant was a test (that would pass *Daubert*) that identifies whether a person had a generally good memory or not. The question then becomes whether that test should be admissible regarding a witness. Some said no, under the theory that such a test would impermissibly invade the role of the jury. Someone asked why not: "I can ask the witness how well he sees without his glasses. I can ask the witness how dark it was. I can do all kinds of things to show the difficulties that a witness in his circumstance would likely have in remembering. Why shouldn't I be allowed to bring in physiological evidence, granted imperfect, that says people with his particular [brain activity] . . . pattern basically don't have very good memories?"

Scientific Evidence:
Frye *and* Daubert

§

The admissibility of scientific information into evidence at trial (for example, an analysis of a brain scan for the purpose of predicting violence) is governed by specific "rules of evidence." The rules of evidence determine what information can be "admitted" (made available for the judge or jury to consider in their decision making), under what conditions, and for what purposes. While each state has its own evidentiary rules, and the federal courts have the Federal Rules of Evidence, there are generally two approaches to the admissibility of scientific evidence. With either set of rules, it is presumed that the theory, data, or technique being offered for admission into evidence has already been determined to be relevant—that is, that the offered information, if reliable, would assist the trier of fact in deciding the instant issue.

The older of the two standards is known as the *Frye* test, after the case in which it was used.[†] The *Frye* test allows for the admission of scientific evidence when the scientific technique, data, or method is "generally accepted" by the scientific community in the relevant field. This test used to be the majority approach in the United States. The courts relied on the members of the relevant scientific discipline for the standard—"general acceptance" was generally proven through additional expert testimony, the citing of standard reference materials in the discipline, and other methods.

[†]   *Frye v. United States*, 293 F. 1013 (D.C. Cir., 1923).

The newer approach, and the one now codified in the Federal Rules of Evidence, is the *Daubert* standard.[‡] The U.S. Supreme Court is the ultimate interpreter of the Federal Rules of Evidence, and with the *Daubert, Joiner*, and *Kumho* rulings, the Court held that federal judges themselves would serve as evidentiary "gatekeepers" and would themselves evaluate the scientific validity of evidence in consideration. In *Daubert*, the Court laid out four criteria for courts to use in their evaluation:

(1) Falsifiability: Can the theory or technique be (and has it been) tested? Drawing on the theories of Karl Popper, the Court pointed out that scientific theories and methods must be falsifiable.

(2) Peer review: Has the theory or technique been subjected to peer review and publication? The Court was sensitive to the fact that some theories are too new or of too limited interest to be published, but it also held that publication and peer review were relevant considerations in assessing validity.

(3) Error rate: What is the known or potential error rate of the methodology or technique?

(4) General acceptance: A *Frye*-like test that considers the degree of acceptance within a relevant scientific community.

The Court held that these four factors formed a flexible rule, and one whose focus should be on determining scientific validity, and therefore evidentiary relevance and reliability. *Daubert* is now the approach taken by the majority of states, and is generally viewed as facilitating the admittance of new science into evidence, since "general acceptance" is no longer the sole test for admissibility.

[‡] The cases in which the United States Supreme Court laid out the current federal standards are *Daubert v. Merrell Dow Pharmaceuticals, Inc.*, 509 U.S. 579 (1993); *General Electric Co. v. Joiner*, 522 U.S. 136 (1997); and *Kumho Tire, Ltd., v. Carmichael*, 526 U.S. 137 (1999). These cases are sometimes referred to collectively as the *Daubert* or *Daubert Joiner Kumho* rules or standards.

Intellectual Property Issues

In general, neuroscience seems unlikely to pose any new challenges for intellectual property law. To the extent that concern may be warranted, it relates to the possibility that neuroscience patents could restrict both the development of "downstream" or derivative products and the development of research tools using the patented devices, processes, or components.

A patent is a type of property right granted to an inventor, giving the exclusive right to the use of some invention (including, for example, mechanical devices, chemical compounds, and manufacturing processes) for a limited period of time. The underlying rationale is that patents reward invention by allowing inventors to exclusively profit from the fruits of their labors and thus making it more likely that inventors will assume the costs and risks of research and invention. The larger society benefits not just from the inventions but also because the inventor is required, as part of the patenting arrangement, to disclose the details of the invention, allowing others to improve, modify, and otherwise build on it.

While an inventor (or patent holder) has exclusive rights to the use of his patented invention, researchers have historically used patented devices or materials (without permission from the patent holder) under a "research exemption." The idea of the research exemption is that researchers may use the patented device or material as long as they are doing so strictly for experimental, research, or other noncommercial purposes. Again, the underlying assumption is that such uses benefit society as a whole (by advancing basic knowledge) but do not compete with the commercial interests of the patent holder.

In light of the recent federal court decision in *Madey v. Duke* (see glossary for a summary), some apprehension exists that the research exemption has been removed. Without a research exemption, scientists may be unable to use the patented devices or materials, or they might have to pay fees that will make use of the patented items prohibitive. The concern that arises is whether such restrictions will inhibit progress in the field by limiting access to the patented tools needed to conduct research or to further innovation. The idea that patents could slow the development of neuroscience, or change the direction in which neuroscience develops, is not trivial. The research exemption does seem to be narrowing, and the reality is that all scientific endeavors, and not just neuroscience, will have to address this issue.

5

Future Directions

THE NEED FOR INCREASED INTERACTION among the legal and neuroscientific disciplines is apparent. One useful form would be scientific educational efforts aimed at lawyers and judges. The Federal Judicial Center's scientific manual[39] is an excellent example of successful targeted scientific education. Similarly, neuroscientists could benefit from education in the legal system's use of science and the types of uses lawyers are foreseeing for neuroscience. Also, the establishment of a formal body, an ongoing conference, or some other mechanism to allow lawyers and neuroscientists to inform each other's work could be quite valuable.

In a related vein, the two communities might cooperate in establishing which neuroscience methods are legally useful and scientifically sound. While the law will probably incorporate new neuroscientific knowledge successfully, it is less clear how that might best occur. A number of participants expressed confidence in the federal *Daubert* standard and similar state-level rules. Proponents of this view believed that judges could determine good science if they simply "did the

---

[39] The Federal Judicial Center (FJC) is the research and education agency of the federal judicial system. The FJC publishes the *Reference Manual on Scientific Evidence,* which is intended to assist federal judges in managing issues regarding scientific evidence, including "recognizing the characteristics and reasoning of 'science' as it is relevant in litigation." See *Reference Manual on Scientific Evidence,* 2d ed., Washington, D.C.: Federal Judicial Center, 2000, p. v.

homework" by preparing adequately and requiring sufficient information from counsel.

However, other participants offered a note of caution: neuroscience is sufficiently complex that *Daubert*-type determinations of admissibility and usefulness may be unduly burdensome on the courts, particularly if the science is introduced to address core jurisprudential issues, such as responsibility. The complexity of neuroscience, and its potential for broad implications, may lead to very lengthy and involved *Daubert* challenges. In addition, there are concerns that a case-by-case method of developing admissibility standards may prove less than optimal—particularly when there could be a collective effort to work with neuroscientists to develop standards for determining "sound" neuroscience in those instances in which the science has yet to be shown reliable or valid. While neither group of participants who expressed concern about admissibility proposed a detailed alternative to *Daubert*, several suggested considering an accrediting process (for labs and technologies), or some legislatively driven creation of an approval process—however, no specific methodology or body was suggested.[40]

Even without a specific strategy for determining which neuroscience methods are legally and scientifically sound, several areas of neuroscientific inquiry are likely to yield useful knowledge for legal proceedings, in addition to their societal impact. These are addiction, enhancement,[41] risk of discrimination, lie detection, and the prediction of behavior.

---

[40] In presenting his paper to the group, Morse did suggest some criteria that would have to be present to consider any type of concerted effort for law reform. The science that would influence legislation would have to be valid and relevant to the legal issue at hand. It would have to have very specific, identifiable implications. And finally, the benefits would have to outweigh the foreseeable costs of reform. Based on these criteria, Morse argued that no likely doctrinal changes in public or private law are on the horizon. There would be a host of civil liberties issues, and the participants' discussion (as well as this report) seems to reflect that assessment.

[41] Particularly of concern to participants were pharmacological enhancements (Ritalin, Provigil) that are already available. As one participant said, " . . . The notion that we are in the future faced with some kind of problem may, in fact, not be correct. We may have the problem now, and what we're talking about is a different manifestation of the current and existing problem."

Finally, one lawyer made the case for expanded clinical testing of the neuroscience technologies likely to be used in legal settings:

> You should be able to do pretty good controlled-clinical-trial kinds of experiments to see whether these things work, whether they work for everybody, whether they work for only certain people, whether you can beat it . . .
>
> You want to put a new drug out, the FDA requires you to go through years and years of detailed clinical trials. There's no such requirement for nonmedical technologies. Does anybody have the interest, the funds, and the will to fund serious rigorous clinical testing of these technologies? If the answer is no, I'd suggest the answer should be changed to yes.

While these sentiments refer specifically to lie detection technologies, opinions like it echoed throughout the meeting regarding many of the technologies discussed. The legal community, while recognizing that many of the relevant neuroscience developments are in their earliest stages of exploration, felt that the scientific community should consider explicit clinical testing of neuroscience tests and technologies for courtroom and other legal uses.[42]

Conclusion

> *Well, there are a lot of things in here that I think are relevant, and there are a lot of things in here that I think are going to scare people and make people misunderstand how much we should pay attention to neuroscientific*

---

[42] In general, early scientific research is often considered more "pure" than "applied" research, in part because the initial exploratory work is often (by necessity) primarily descriptive and explanatory, rather than an attempt to manipulate or alter the phenomenon or mechanism of interest. In the case of neuroscience, many of the applied uses are only beginning to be developed. Here, the lawyers were essentially arguing that since we expect the scientific knowledge and technologies to be used in legal settings, perhaps researchers should conduct some experimentation directly addressing the potential legal uses.

evidence in the courts, and we have to be very careful to
distinguish those.

—Neuroscientist participant, on the risks
of imprecision in the way we portray the
understandings of neuroscience

A concern strongly expressed at the meeting was that both lawyers and neuroscientists be cautious about how the science is used and presented. For the well-being of both fields, the science must be presented, used, and discussed in a realistic and accurate fashion—one that reflects both the limitations and the potentials of the science. As one participant put it, it is time for neuroscientists to start identifying and delineating the boundaries of what is known and likely to be knowable—the limits of neuroscience knowledge. This will enable the legal community to better appreciate what neuroscience can and cannot tell us and to what uses neuroscience can be put in the service of the law, and of society.

At the same time, that the future is not fully knowable is not reason to delay the dialogue. As one neuroscientist noted, "We really do have an obligation to think about things, even if they don't seem likely right now, because they will come faster than we can possibly believe."

PART II

Commissioned Papers

Free Will in the Twenty-first Century

A Discussion of Neuroscience and the Law

MICHAEL S. GAZZANIGA

Center for Cognitive Neuroscience,
Dartmouth College, Hanover, New Hampshire

and

MEGAN S. STEVEN

University Laboratory of Physiology, University of Oxford, Oxford, UK

IMAGINE YOU ARE A JUROR for a horrific murder case. You are well informed and know some things about the American criminal justice system. First, you are aware that 95 percent of criminal cases never come to trial. Most cases are either dismissed or plea-bargained. Second, you understand that there is a larger probability that your defendant is guilty than that he is innocent.

So you sit down with 11 of your peers, people who may not be up on the latest scientific understanding about human behavior. Indeed, you know most jurors don't buy criminal-defense arguments dealing with such things as "temporary insanity," and many consider these arguments excuses. Jurors are tough, practical people. That is the profile of the American jury system: nothing fancy, just 12 people trying to make sense out of a horrible event. Most of them have never heard of the word *neuroscience* nor given a moment's thought to the concept of "free will." They are there to determine whether the defendant committed the crime, and if they decide that he did, they will deliver their verdict without regret.

It is against this backdrop of life in the American courthouse that we come to consider the perennial question: Do human beings have "free will"? When a jury decides that the defendant did in fact commit the crime, have they considered whether the person did so of his own free will or as an inevitable action based on the nature of his brain and his past experiences? The short answer is no, they have not. As with so many issues in which modern scientific thinking confronts everyday realities, the man on the street is not waiting for the answer. Yet the contention of this paper is that it is important to examine these issues because someday—a day not so far away—they will dominate the legal community.

Currently, brain mechanisms are being discovered that are helping us understand the role of genes in building our brains, the role of neuronal systems in allowing us to sense our environment, and the role of experience in guiding our future actions. It is now commonly understood that changes in our brain are both necessary and sufficient for changes in our mind. Indeed, an entire subfield of neuroscience (called cognitive neuroscience) has arisen in recent years to study the mechanisms of this occurrence.

With this reality of twenty-first-century brain science, many people find themselves worrying about that old chestnut free will. The logic goes like this: The brain determines the mind, and the brain is a physical entity subject to all the rules of the physical world. The physical world is determined, so our brains must also be determined. If our brains are determined, and if the brain is the necessary and sufficient organ that enables the mind, then we are left with the questions, Are the thoughts in our mind that arise from our brain also determined? Is the free will we seem to experience just an illusion? And if free will is an illusion, must we revise our concepts of what it means to be personally responsible for our actions?

This dilemma has haunted philosophers for decades. But with the advent of neuroimaging tools that allow scientists a noninvasive means to study the human brain in action, these questions are now being explored by neuroscientists, and, increasingly, the legal world is seeking answers. Defense lawyers are looking for that one pixel in their client's brain scan that shows an abnormality, a predisposition to crime or a malfunction in normal inhibitory networks, thereby allowing for the argument, "Harry didn't do it. His brain did it. Harry is not responsible for his actions."

Although neuroscience will offer us some answers, ultimately we must realize that even if the causation of an act (criminal or otherwise) is explainable in terms of brain function, this does not mean that the person who carries out the act is exculpable. This essay presents a brief summary of current ideas on free will from both a philosophical and a legal point of view, followed by a discussion of the neuroscience evidence that suggests that our brains drive our actions but not our personhood.

Our own perspective will then be presented, with some ideas on the implications of current neuroscience understanding. In brief, we will argue that while brains can be viewed as more or less automatic devices, like clocks, we as people are usually free to choose our own destiny.

The Philosophical Stance on Free Will

INDETERMINISM

One of the earliest discussions of free will came from Epicurus, who was a complete physicalist.[1] He understood free will as occurring when atoms swerve in an undetermined way. Other attempts to understand free will were made by dualists, philosophers who believed there was a difference between mind and body, that in fact the mind and brain were completely different entities. This view was widely referred to as the "ghost in the machine."[2] According to the famous proponent of dualism René

---

[1] Epicurus, "Letter to Herodotus."

[2] Ryle 1949.

Descartes, the human body is a machine, and only in combination with a metaphysical soul does it become a person.[3] Free will, thus, comes from the metaphysical "ghost" and not from the "machine" (the brain).

Philosophers call the denial of determinism "indeterminism." Dualists need not be determinists, and, as Epicurus shows, indeterminists need not be dualists. It is, admittedly, easier to defend indeterminism if one is a dualist instead of a physicalist, and indeterminists are usually dualists.

Even though modern science finds the idea of dualism untenable, it reigns as the predominant perception of the self. Most dualists endorse the concept of free will and the notion that we are all practical reasoners driven by "free will" to make rational choices in our behavior. This way of thinking largely influences modern law. Still, as we will see later, one need not be a proponent of dualism to argue for the view that we are free to make our own decisions.

DETERMINISM

In stark contrast to indeterminism is determinism, the philosophical doctrine that every act, action, or situation, human or otherwise, is inevitable. In the common view, if determinism holds, free will is only an illusion, for how could an act be freely initiated if every act is inevitable?

It is important to distinguish inevitability from probability when defining determinism. Consider quantum mechanics. Some interpretations of quantum mechanics imply that some events (such as electrons jumping energy levels) are merely probable, not inevitable. (This is a modern version of Epicurus' swerve theory.) We do not think the extension of this view to human behavior is useful to the present discussion. Arbitrariness cannot lead to voluntary actions or responsibility any more than determinism can. Voluntary acts must be freely initiated, not just be the result of randomness or probability. Thus, the argument we present against determinism could just as easily be an argument against arbitrariness or probability.

---

[3] Descartes 1641.

Yet, the question arises, If determinism is true, what is it that is doing the determining? Traditionally, genes have been implicated as the predictors of our destiny. Stephen Jay Gould, by no means an advocate for the idea of genetic determinism, explained the theory: "If we are programmed to be what we are [by our genes], then [our] traits are ineluctable. We may, at best, channel them, but we cannot change them either by will, education or culture."[4] Indeed, some processes are largely determined by our genes (for example, if someone has the gene for Huntington's disease, he or she will almost certainly contract the disease, and "good living, good medicine, healthy food, loving families or great riches can do nothing about it"[5]), but in reality, many of our traits are not entirely coded in our genes. Our environment and chance also play a role in determining our traits and behaviors.

While genes build our brains, it is our brains, actively making millions of decisions per second, that ultimately enable our cognition and behavior. Now we'll examine the arguments in support of neural determinism (the determinism of the brain) by describing them in two ways: generally, and then by means of the specific example of violent criminal behavior.

General Arguments Supporting Free Will

If the brain carries out its work before one becomes consciously aware of a thought, it would appear that the brain enables the mind. This is the underlying idea behind the neuroscience of determinism. The argument came to everyone's attention with the work of Benjamin Libet, in the 1980s.[6]

Libet measured brain activity during voluntary hand movements. He found that between 500 and 1,000 milliseconds (ms) before we actually move our hand there is a wave of brain activity, the readiness potential.

---

[4] Gould 1997, p. 238.

[5] Dennett 2003, p. 157.

[6] See Libet 1991 for a review. See also Libet, Benjamin, *Mind Time: The Temporal Factor in Consciousness*, Cambridge, MA: Harvard U. Press, 2004.

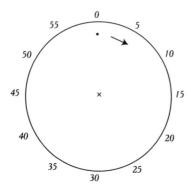

Figure 1. *Adapted from Libet 1999, p. 45.*

Libet set out to determine "the notorious 'time t,'"[7] somewhere in that 500–1,000 ms, when we make the *conscious decision* to move our hand.

Libet measured the activity of his subjects' brains using a technique known as event-related potentials (ERP) while the subjects were told to make a conscious and voluntary hand movement. The subject would stare at a clock (see Figure 1), and at the very moment he made the conscious decision to flick his wrist, he would remember the position of the moving black dot (for example, at the 5 or 53 mark).

After the movement was made, the subject would report to the experimenter the position of the dot at the moment he made the conscious decision to move his hand. Libet would then correlate that moment with the time that a readiness potential from the subject's brain waves was recorded.

Libet found that even before "time t," when the subject first became consciously aware of his decision to make a hand movement, the subject's brain was active—the readiness potential was present. The time between the onset of the readiness potential and the moment of conscious decisionmaking was about 300 milliseconds. If the readiness potential of the brain is initiated before we are aware of making the decision to move our hand, it would appear that our brains know our decisions before we ever become conscious of them.

---

[7] Dennett 2003, p. 228.

The evidence thus seemed to favor the illusion and not the actuality of free will. Interestingly, Libet argued that since the time from onset of the readiness potential to the actual hand movement is about 500 ms, and it takes 50–100 ms for the neural signal to travel from the brain to the hand to actually make it move, then there are 100 ms left for the conscious self to either act on the unconscious decision or veto it. That, he said, is where free will comes in—in the vetoing power. Vilayanur Ramachandran, in an argument similar to John Locke's theory of free will,[8] suggests that "our conscious minds may not have free will but rather 'free won't' "![9] (For a critique of Libet's experiments, see Dennett 2003, pages 228–231.)

Platt and Glimcher at New York University study the monkey brain and examine the activity of neurons in a part of the brain called the inferior parietal lobule. Their experiments fortify the notion that the brain acts on its own before we become consciously aware of its actions. Each neuron in the inferior parietal lobule has a receptive field that prefers one area of the visual world over all others. For example, when a monkey fixates on a central location and the experimenter moves a bar of light around on the wall, there will be a single place—say, five inches over and five inches up from the fixated spot—where a particular neuron sends, or "fires," signals at a higher rate than anywhere else. Any time the bar of light is in that zone, the cell fires signals like a machine gun, and it stops firing when the bar of light is outside the zone, that is, outside the neuron's receptive field.

Through a long series of experiments, Platt and Glimcher showed that these neurons know a great deal about their receptive fields. They "know" (or respond to changes in) the probability that a light in their field is the target light; they even "know" the probability of getting a reward (usually a sip of juice for a monkey)—the cells demonstrate this knowledge by firing more when there is a greater probability of payoff.[10] Platt and Glimcher also found that neurons

---

8 Locke 1690. Book II, chapter XXI, paragraph 47.
9 Ramachandran 1998.
10 Platt and Glimcher 1999.

in this area of the brain change their firing patterns depending on the size of the reward likely to be received. This is some of the first work to reveal that the parietal lobe plays a critical role in deciding whether or not to act. The neurons in this part of the brain are not passively attending to a specific part of the visual world; rather, they appear to be tied to the *utility* of the movement and may actively aid in deciding whether to make a movement. All of this happens long before there is even a hint that the animal is deciding what to do. The automatic brain is at work.

Many experiments underline how the brain gets things done before we know about it. Another example is derived from research by one of the present authors.[11] Our brains are wired so that if we fixate visually on a point in space, everything to the right of the point is projected to the visual areas in the left side of the brain, and everything to the left of the fixation point is projected to the visual areas in the right side of the brain. The two sides of the brain, however, are interconnected through a large fiber tract called the corpus callosum.

Experiments have shown that when the word *he* is presented to the left of the fixation point and the word *art* is presented to the right of the fixation point, the subject perceives the word *heart*. This integration is achieved without the subject's conscious awareness. Electrophysiological recordings, carried out in collaboration with Ron Mangun and Steven Hillyard, helped decipher how this is accomplished in the brain, and shed light on another example of how the brain acts and makes decisions well before we are aware of our integration of *he* with *art*.

Electrical potentials in the brain evoked by stimuli presented to a subject's senses can be measured by the previously mentioned technique of event-related potentials. This procedure enables one to track, over time, the activation pattern of neurons in the cortex of one side of the brain and their cross-hemispheric connections through the corpus callosum to the opposite side of the brain. Gazzaniga, Mangun, and Hillyard found that after a stimulus (for example, *he*) is presented to the

---

[11]   Gazzaniga and LeDoux 1978.

left visual field, the right visual cortex quickly activates (and the converse is true for stimuli presented to the right visual field). About 40 ms later, the activity begins to spread to the opposite hemisphere; after another 40 ms or so, the information arrives to consciousness and—voilà—*heart* appears. The integration of *he* and *art* occurs long before we become consciously aware of the output *heart*.

FREE WILL AND VIOLENCE

Now that we have outlined some of the general arguments in neuroscience relating to the concept of free will—namely, that the brain makes many decisions before we are aware of them—let us consider a real-life problem of free will: violent criminal behavior. Specifically, let us discuss how those interested in the neurological mechanisms of violence could use current neuroscientific knowledge to argue for reduced culpability under the law, and why they would usually be wrong to do so.

People who commit repeated violent crimes often have antisocial personality disorder (APD), a condition characterized by deceitfulness, impulsivity, aggressiveness, and lack of remorse.[12] People with APD have abnormal social behavior and lack the inhibitory mechanisms often associated with normal frontal lobe functioning. Ever since the famous case of Phineas Gage in 1848, psychologists have known of the critical role played by the frontal lobe in normal social behavior. Without it, it seems the capacity to use "free won't" is impaired.

Phineas Gage is one of the most well known neuropsychological patients of all time. He survived an explosion during the construction of a railroad that drove an iron bar through his head, damaging parts of the frontal areas of his brain.[13] Following his recovery from the accident he appeared normal, but those who had known him before the accident noticed some changes. His friends said that Gage was "no longer Gage."[14]

---

[12] See the *Diagnostic and Statistical Manual of Mental Disorders IV* diagnostic criteria for Antisocial Personality Disorder.

[13] Nolte 2002, p. 548.

[14] Harlow 1868, p. 327.

Indeed, his personality changed drastically after the accident. He was impulsive, lacked normal inhibitions, and demonstrated inappropriate social behaviors (for example, he would swear and discuss things of a sexual nature in situations where it was not appropriate to do so). Unfortunately, no autopsy was done to determine the precise injury to the frontal part of Gage's brain, but contemporary reconstruction based on the skull damage indicates that the lesion was localized to medial and orbital regions of the prefrontal cortex.[15]

Evidence from other patients with lesions of the prefrontal regions of the brain confirms these findings.[16] The question then arises: Do criminals with APD, who demonstrate abnormal social behaviors similar to those of patients with prefrontal lobe damage, have abnormalities in the prefrontal areas of their brain? To address this question, Raine et al. (2000) imaged the brains of 21 people with APD and compared them with the brains of two control groups: healthy subjects and subjects with substance dependence. Because substance dependence often accompanies APD, the experimenters wanted to ensure that they could discern the brain differences due solely to the APD and not to the substance abuse.[17] The researchers found that people with APD had a reduced volume of gray matter and a reduced amount of autonomic activity in the prefrontal areas of their brain as compared with both control groups. The findings indicate that there is a structural difference (that is, in the amount of gray matter in the prefrontal lobe relative to the rest of the brain) between the brains of criminals with APD and the brains of the normal population. This also suggests that a volume difference in gray matter may lead to a functional difference in social behavior between the two groups.

---

[15] See Damasio et al. 1994 for the details of that reconstruction.

[16] Damasio 2000.

[17] It should be noted that there are some critics of the findings of this study who question whether Raine et al. were really able to weed out the effects of substance abuse with their control group. These critics suggest that a more cautious conclusion from the study would be that "APD combined with [substance use] is associated with prefrontal cortical volume reductions." See Seifritz et al. 2001 for more on this critique and for a response from Raine et al.

Further support for this theory comes from the case study of a boy who displayed characteristics of APD from a young age. When playing Russian roulette, he shot himself in the head and damaged his medial prefrontal cortex.[18] Amazingly, he survived the injury. Those who knew him well before the brain damage report that there was little or no change in his personality after it. Behavioral characteristics before his injury suggest that the boy's medial prefrontal cortex was not working properly (possibly due to a reduced volume of gray matter), and the continuation of those behavioral problems after his injury indicate that damage to the already malfunctioning medial prefrontal cortex had little or no effect on his behavior.

A study by New et al. in 2002 looked at a specific characteristic of APD—impulsive aggression. Using a brain imaging technique called positron-emission tomography (PET), they measured the metabolic activity of the brain in response to a serotonergic (involving serotonin, an excitatory brain chemical) stimulus called m-CPP in various areas of the brain in people with impulsive aggression, and then compared the images with the scans of healthy, nonaggressive controls. M-CPP normally activates the anterior cingulate (a frontal area of the brain known to be involved in inhibition and suspected of having been damaged in Phineas Gage's brain) and deactivates the posterior cingulate. The opposite was found to be true of people with impulsive aggression—that is, the anterior cingulate was deactivated and the posterior cingulate was activated in response to m-CPP. The investigators concluded that patients with impulsive aggression exhibit a decreased activation of inhibitory regions (for example, in the anterior cingulate) in response to serotonergic stimulus and that this may contribute to their difficulty in modulating aggression.[19]

If true, it is possible that these people merely do not inhibit their impulses, even though they *could* inhibit them (and therefore should be held responsible for their actions). Future research will be needed to determine how much prefrontal damage or gray matter loss is necessary

---

[18]   See Bigler 2001.

[19]   See New et al. 2002, p. 621.

for the cessation of inhibitory function in the brain (and thus perhaps for the mitigation of responsibility). Neuroscientists must realize, however, that when considering the specific example of violence and the brain, the argument might be made that the correlation of any given brain state with nonviolent behavior could be just as high as its correlation with violent behavior. Specifically, most patients who suffer from Gage-type lesions involving the inferior orbital frontal lobe do not exhibit antisocial behavior of the sort that would be noticed by the law. Even though a patient's spouse may be able to sense changes in his or her behavior, the patient is still constrained by all the other forces in society, and the frequency of his or her abnormal behavior is no different than would be seen in the normal population. The same is true for people with schizophrenia. The rate of aggressive criminal behavior is not greater among schizophrenics than it is among the normal population. Since people with Gage-type lesions or with schizophrenia are no more likely to commit violent crimes than are normal people, it seems that merely having one of these brain disorders is not enough to remove responsibility.

These facts, however, leave pure determinists unimpressed. Since their position is based on a theory of causation, and since they believe all actions have definable inputs, they also believe that our current shortcomings in understanding causation are just that—shortcomings. We will argue, however, that there is evidence for the existence of free will in a deterministic world.

Free Will in a Deterministic System

Although mechanistic descriptions of how the physical brain carries out behavior have added fuel to determinism, philosophers and scientists alike have argued that the concept of free will can coexist, even if one assumes that the brain is as mechanical as clockwork. These views challenge the idea that explanation of mechanism leads to exculpation.

In 1954, A. J. Ayer put forth a theory of "soft determinism" (or the conjunction of determinism and compatibilism—the belief that determinism is compatible with free will). He argued, as had many philosophers, including David Hume, that even in a deterministic world, a person can still act freely. Ayer posits that free actions are actions that result from desires, intentions, and decisions without external compulsion or constraint. He distinguishes between free actions and constrained actions (not between uncaused and caused actions). Free actions are those that are caused by internal sources, by one's own will (unless one is suffering from a disorder), whereas constrained actions are those that are caused by external sources (for example, by someone or something forcing you physically or mentally to perform an action, by hypnosis, or by disorders like kleptomania). When someone performs a free action to do A, he or she could have done B. When someone makes a constrained action to do A, he or she could have done only A.[20] Thus, Ayer argues that actions are free as long as they are not constrained. Free actions are dependent not on the existence of a cause but on the source of the cause. Though Ayer does not explicitly discuss the brain's role, one could make the analogy that those actions—and indeed those wills—that are caused by, say, a disease-free brain or hypnosis are not constrained (even though they may be determined). In this way, the brain is determined, but the person is free.

More recently, Daniel Dennett has produced another argument and demonstration that free will can exist in a deterministic system. He describes a model of determinism and proves that the model allows for free will, thereby implying that actual determinism might also allow it. Dennett's model consists of a set of deterministic worlds, called life worlds. These life worlds are based on simplistic rules that help one visualize determinism and evolution. Basically, a life world is a grid of pixels (some filled in, or ON, and some empty, or OFF). Each pixel has eight surrounding pixels, and the only rules of the life world are as follows: "For each cell in the grid, count how many of its eight neighbors is [*sic*] ON at the present instant. If the answer is exactly two, the cell stays in its present state (ON or OFF) in the next instant. If the answer

---

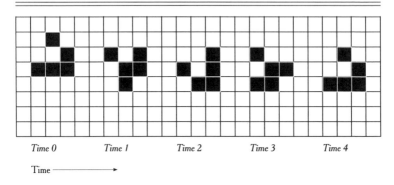

Time 0   Time 1   Time 2   Time 3   Time 4

Time ——————→

Figure 2. *Example of a glider. Adapted from Dennett 2003.*

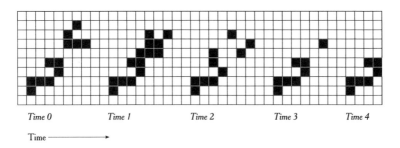

Time 0   Time 1   Time 2   Time 3   Time 4

Time ——————→

Figure 3. *Example of an eater eating from a glider. Adapted from Dennett 2003.*

is exactly three, the cell is ON in the next instant whatever its current state. Under all other conditions the cell is OFF."[21]

The life world is thus deterministic—there is no escaping the rules, and every future moment is determined by the arrangement of pixels in the past. But by studying the shapes and patterns of the evolving ON and OFF pixels, Dennett has discovered that some patterns glide across the plane but keep their shape over time (he calls these gliders—see Figure 2); some shapes are eaters (see Figure 3) and can actually consume the gliders. There are other configurations that can avoid the eaters.[22]

---

[21] Dennett 2003, p. 36.

[22] Paraphrased from Dennett 2003, pp. 38–45.

It logically follows that:

In some deterministic worlds there are avoiders avoiding harms.

Therefore, in some deterministic worlds some things are avoided.

Whatever is avoided is avoidable, or evitable.

Therefore, in some deterministic worlds not everything is inevitable.

Therefore, determinism does not imply inevitability.[23]

Thus, "if determinism is true, then whatever happens is the determined [not inevitable] outcome of the complete set of causes that obtain at each moment."[24] Indeed, "many thinkers assume that determinism implies inevitability. It doesn't."[25] If things can be avoidable in a deterministic world, then Dennett believes that determinism does not exclude the possibility of a free will.

Automatic Brains and Interpretive Minds

Views such as those outlined above all cleverly address the issue of free will, though many remain controversial. In our own view, ghosts in the machine, emergent properties of complex systems, logical indeterminacies, and other characterizations miss the fundamental point. We conclude that brains are automatic but people are free.

With each passing decade, more is known about the mechanistic action of the nervous system and how it produces perceptual, attentional, and mnemonic function and decisions. While this assertion serves as the motivating aspiration of modern neuroscience, it should also be observed that we still have much to learn about how the brain enables the mind.

We recently attended a conference at which more than 80 leading scientists presented their research findings on how the brain enables the mind. It soon became obvious that the central question of cognitive neuroscience (the enabling of the mind by the brain, or understanding what is driving the mind) remains not only unanswered

---

[23] Dennett 2003, p. 56.

[24] Ibid., p. 57.

[25] Ibid., p. 25.

but—worse—unexamined. The brain scientists who are addressing issues of human cognition are producing work that illuminates which brain systems correlate with particular measurable human behaviors. For example, a series of studies might investigate which areas of the visual system become activated when a subject attends to a particular visual stimulus. While these discoverable correlations are of interest, the question of how the brain knows whether, when, and how to increase the gain of a particular neuronal system remains unanswered. Overall, modern studies always seem to leave room for the homunculus, the little ghost in the machine, which does all the directing of brain traffic. It is commonplace to hear the phrase "top-down processes versus bottom-up processes" (that is, processes driven by feedback from "higher" areas of the brain rather than direct input from sensory stimuli), but the fact is that no one knows anything about the "top" in "top-down." This is a major problem of cognitive neuroscience today, and we hope that it will become the subject of research in the near future.

Rather than lamenting any further the gaps in our understanding, let us move on to what we do know about the brain and how that knowledge can influence the law. To address this, we must have in mind the current legal system's view of human decision making, and to that end we will consider Harry and his crime of murder.

Under our legal system, a crime has two defining elements: the *actus reus*, or proscribed act, and the *mens rea*, or guilty mind. For Harry to go to jail, both elements have to be proven beyond a reasonable doubt. In general terms, the legal system works hard to determine the agency of the crime. Where it wants help from neuroscience is on whether or not Harry should be held "personally responsible" for the crime. Did Harry do it, or did his brain? This is where the slippery slope begins. Our argument is that neuroscience can offer very little to the understanding of responsibility. Responsibility is a human construct, and no pixel on a brain scan will ever be able to show culpability or nonculpability.

In practice, legal authorities have had great difficulty crafting standards to divide the responsible from the nonresponsible. The various rules for a finding of legal insanity, from the 1843 M'Naghten decision to the twentieth-century Durham and ALI Model Penal Code tests, have all been found lacking.[26] Experts for the defense and prosecution argue different points from the same data. The idea here is to have neuroscience come to the rescue of the current scheme.

At the crux of the problem is the legal system's view of human behavior. It assumes Harry is a "practical reasoner," a person who acts because he has freely chosen to act. This simple but powerful assumption drives the entire legal system. Even though we all might conceive of reasons to contravene the law, we can decide, because we have free will, not to act on such thoughts. If a defense lawyer can provide evidence that a defendant had a "defect in reasoning" that led to his inability to stop from committing the crime, then the defendant, Harry, can be exculpable. The legal authorities want a brain image, a neurotransmitter assay, or something to show beyond a reasonable doubt that Harry was not thinking clearly, indeed *could* not think clearly and stop his behavior.

The view of human behavior offered by neuroscience is simply at odds with this idea. In many ways it is a tougher view, and in many other ways it is more lenient. Fundamentally, however, it is different. Putting aside the caveats mentioned above, neuroscience is in the business of determining the mechanistic actions of the nervous system. The brain is an evolved system, a decision-making device that interacts with its environment in a way that allows it to learn rules to govern how it responds. It is a rule-based device that, fortunately, works automatically. Because this kind of formulation usually elicits howls, let us quote from one of the authors' earlier works:

> "But," some might say, "aren't you saying that people are basically robots? That the brain is a clock, and you can't hold people responsible for criminal behavior any more than you can blame a clock for not working?" In a word, no. The comparison is inappropriate; the issue (indeed, the very notion) of responsibility has not emerged. The

---

[26] Waldbauer and Gazzaniga 2001.

neuroscientist cannot talk about the brain's culpability any more than the watchmaker can blame the clock. Responsibility has not been denied; it is simply absent from the neuroscientific description of human behavior. Its absence is a direct result of treating the brain as an automatic machine. We do not call clocks responsible precisely because they are, to us, automatic machines. But we do have other ways of treating people that admit judgments of responsibility—we can call them practical reasoners. Just because responsibility cannot be assigned to clocks does not mean it cannot be ascribed to people. In this sense human beings are special and different from clocks and robots.[27]

This is a fundamental point. Neuroscience will never find the brain correlate of responsibility, because that is something we ascribe to humans, not to brains. It is a moral value we demand of our fellow, rule-following human beings. Just as optometrists can tell us how much vision a person has (20/20, 20/40, or 20/200) but cannot tell us when someone is legally blind or has too little vision to drive a school bus, so psychiatrists and brain scientists might be able to tell us what someone's mental state or brain condition is without being able to tell us (unless arbitrarily) when someone has too little control to be held responsible. The issue of responsibility (like the issue of who can drive school buses) is a social choice. According to neuroscience, no one person is more or less responsible than any other for actions. We are all part of a deterministic system that some day, in theory, we will completely understand. Yet the idea of responsibility is a social construct and exists in the rules of the society. It does not exist in the neuronal structure of the brain.

Summary

Issues surrounding the study of neuroscience and the law are compelling and absorbing. It would be rash to conclude on any other note than one of modesty about our current understanding of the problems discussed in this paper. Much work is needed to further clarify the complex issues raised by both the law and the brain sciences.

---

[27] Waldbauer and Gazzaniga 2001.

Still, we would like to offer the following axiom based on modern understanding of neuroscience and on the assumptions of legal concepts: Brains are automatic, rule-governed, determined devices, while people are personally responsible agents free to make their own decisions. Just as traffic is what happens when physically determined cars interact, responsibility is what happens when people interact. Brains are determined; people are free.

Acknowledgments

We are in debt to our colleague Walter Sinnott-Armstrong for many fine suggestions regarding clarity and accuracy.

Works Cited

Ayer, A. J. (1954). "Freedom and Necessity." In A. J. Ayer, ed., *Philosophical Essays.* London: Macmillan.

Bigler, E. D. (2001). "Frontal Lobe Pathology and Antisocial Personality Disorder." *Archives of General Psychiatry* 58: 609–611.

Damasio, A. R. (2000). "A Neural Basis for Sociopathy." *Archives of General Psychiatry* 57: 128.

Damasio, H., et al. (1994). "The Return of Phineas Gage: Clues About the Brain from the Skull of a Famous Patient." *Science* 264 (5162): 1102–1105.

Dennett, D. C. (2003). *Freedom Evolves.* New York: Viking Press.

Descartes, R. (1641). *Meditations Metaphysiques.*

Diagnostic and Statistical Manual of Mental Disorders, 4th ed. (1994). Washington, D.C.: American Psychiatric Association.

Epicurus (c. 300 B.C.). "Letter to Herodotus." Available online at www.epicurus.net/herodotus.html.

Gazzaniga, M. S., and J. E. LeDoux (1978). *The Integrated Mind.* New York: Plenum.

Gould, S. J. (1997). *Ever Since Darwin.* New York: W. W. Norton.

Harlow, H. M. (1868). "Recovery From the Passage of an Iron Bar Through the Head." *Massachusetts Medical Society Publication* 2: 327.

Libet, B. (1991)."Conscious vs. neural time." *Nature* 352 (6330): 27–28.

Libet, B. (1999). "Do we have free will?" *Journal of Consciousness Studies* 6 (8–9): 45.

Locke, J. (1690). *An Essay Concerning Human Understanding*.

New, A. S., et al. (2002). "Blunted Prefrontal Cortical 18Fluorodeoxy-glucose Positron Emission Tomography Response to Meta-Chlorophenylpiperazine in Impulsive Aggression." *Archives of General Psychiatry* 59: 621–629.

Nolte, J. (2002). *The Human Brain: An Introduction to Its Functional Anatomy*, 5th ed., pp. 548–549. St. Louis: Mosby.

Platt, M. I., and P. W. Glimcher (1999). "Neural correlates of decision making in parietal cortex." *Nature* 400: 233–238.

Raine, A. (2000). Comment on Seifritz et al. "Is Prefrontal Cortex Thinning Specific for Antisocial Personality Disorder?" *Archives of General Psychiatry* 58: 402.

Raine, A., et al. (2000). "Reduced Prefrontal Gray Matter Volume and Reduced Autonomic Activity in Antisocial Personality Disorder." *Archives of General Psychiatry* 57: 119–127.

Ramachandran, V. (1998). As quoted in B. Holmes, "Irresistible illusions." *New Scientist* 159, no. 2150 (September 5, 1998): 32.

Ryle, G. (1949). *The Concept of Mind*. London: Hutchinson.

Seifritz, E., et al. (2001). "Is Prefrontal Cortex Thinning Specific for Antisocial Personality Disorder?" *Archives of General Psychiatry* 58: 402.

Waldbauer, J. R., and M. S. Gazzaniga (2001). "The Divergence of Neuroscience and Law." 41 *Jurimetrics* 357 (Symposium Issue).

Neuroscience Developments and the Law

LAURENCE R. TANCREDI
*Clinical professor of psychiatry at New York University School of Medicine.
Dr. Tancredi is also an attorney.*

D EVELOPMENTS IN NEUROSCIENCE are escalating at an ever increasing pace, mimicking in many respects the pattern of development in genetics during the sixties, seventies, and eighties, following the delineation of the genetic code. Imaging technologies such as positron-emission tomography (PET), single-photon-emission computed tomography (SPECT), computerized electroencephalography (CEEG), and functional magnetic resonance imaging (fMRI), to name a few, have been pivotal in this expansion of knowledge. By virtue of their ability to penetrate the physical inaccessibility of the brain, these technologies have permitted for the first time the direct investigation of the functioning brain in a living human being.[1] These observations have contributed greatly to the breakdown of the historical notion that a mind-brain dichotomy operates, that mental things have an intrinsic nature. Advances brought about by these technologies have aided in demonstrating a relationship

---

[1] See L. R. Tancredi and N. D. Volkow. 1992. "A theory of the mind/brain dichotomy with special reference to the contribution of positron emission tomography." *Perspectives in Biology and Medicine* 35, 549–571.

between biological processes in the physical brain and the transmission of information and sensations over neuron pathways.[2] Along with neurochemical and anatomical techniques, as well as neuropsychological studies, they have also identified specific areas in the brain that deal with cognition, perception, and emotions.[3]

Neuroscience advances have far outstripped the ability of social institutions, especially the law, to accommodate their principles. With the rate of new discoveries, it may not be long before technologies will provide a window not only into the process but the very content of the functioning brain. This discussion addresses the possible impact of neuroscience research findings both for the short term and the long term on four broad and closely related areas of brain function that are of legal interest: brain death; cognition as it applies to competency in civil matters; societal aspects of enhancing cognitive functions beyond what is considered normal; and neuroscience measures for determining a person's veracity. The social impact of research developments in these four areas, particularly as they offer the opportunity for discrimination, unfairness, and inequity, will also be examined.

Brain Death

The introduction of the mechanical ventilator into medical practice in the mid-twentieth century was not only a major advance in the care of critically ill patients, but it also introduced the possibility of changing the definition of death from the long-held tradition of cessation of heartbeat to the cessation of brain function.[4] This transformation was in large part fueled by the combination of mechanical ventilation devices and new

---

[2] Physical processes that appear to correlate with processes in the mental realm have been shown by these technologies, though strong biological evidence has not been established thus far to show consistently that when a physical reaction occurs it relates directly to a specific form of mental activity. See footnote 1, pp. 560–567.

[3] Ibid., 549.

[4] M. N. Diringer and E. F. M. Wijdicks. 2001. "Brain Death in Historical Perspective." In E. F. M. Wijdicks, ed., *Brain Death*. Philadelphia: Lippincott, Williams and Wilkins, 5–27.

cardiac stimulation measures, which keep people alive long after anoxia has destroyed their central nervous system. But even more significant than these developments, the feasibility of organ transplantation that had been demonstrated for kidneys and hearts in the 1960s promised to usher in a new era of medical treatment. Brain death became the requisite for organ donation.[5] In general terms, besides requiring the coexistence of apnea (the state of being without respiration) and coma, brain death has been defined in two formulations: the complete and irreversible cessation of all brain function, including that of the brain stem (whole-brain formulation), and the irreversible and complete end of brain-stem function alone.[6]

The Harvard Medical School in 1968 established a committee under the leadership of Dr. Henry Beecher, an anesthesiologist, to define death predicated on neurological criteria.[7] The objective behind this effort was primarily that of assisting physicians to withdraw life-extending treatments, such as mechanical ventilation, in futile cases.[8] In fact, the development of mechanical ventilators was one of the two major advances in health care that influenced the Harvard group to shift to a neurological determination of death. The other development was the creation of intensive care units (ICUs), which kept many people alive who would otherwise have expired.[9] Some of those kept alive were in terminal vegetative states. Facilitating the transplantation of organs was not a consideration by the Harvard committee. "Irreversible coma," or brain death, required a coma of identifiable cause, unresponsiveness or

---

[5] E. F. M. Wijdicks. 2001. "Current concepts: The diagnosis of brain death." *New England Journal of Medicine* 344, 1215–1221.

[6] M. DeTourtchaninoff, P. Hantson, P. Mahieu, and J. M. Guerit. 1999. "Brain death diagnosis in misleading conditions." *QJM* 92, 407–414.

[7] Note: Cerebral function as a basis for defining death among patients on mechanical ventilation was reported as early as 1956, and the term *"le coma depasse"* was used in 1959 in Europe to define death based on neurological criteria. See footnote 5, at 1215.

[8] See footnote 4, p. 13.

[9] S. D. Shemie, C. Djoig, and P. Belitsky. 2003. "Advancing toward a modern death: The path from severe brain injury to neurological determination of death." *CMAJ* 168, 993–995.

lack of receptivity to stimuli, the absence of breathing and movement, and the absence of brain-stem reflexes.[10]

In 1981 the President's Commission for the Study of Ethical Problems in Medicine and Biomedical and Behavioral Research promulgated criteria for neurological death that would essentially eliminate the possibility of error in identifying a living person as dead, and minimize error in determining that a dead person was indeed alive.[11] This report reduced the period of observation necessary to determine death and emphasized the use of confirmatory tests to justify that reduction.[12]

In 1995 the American Academy of Neurology developed guidelines for defining brain death.[13] These guidelines, which have been adopted in many health care institutions, include the following: the establishment of coma; determination of the cause of the coma; absence of confounding factors; absence of brain-stem reflexes and motor responses; and apnea. The guidelines recommend that a repeat evaluation be conducted within six hours of the initial clinical examination. Furthermore, they require confirmatory laboratory tests only when specific elements of the clinical testing cannot be evaluated in a reliable manner.[14]

---

[10]   See 1968 "A definition of irreversible coma: Report of the Ad Hoc Committee of the Harvard Medical School to Examine the Definition of Brain Death." *Journal of the American Medical Association*, 205, 337–340.

[11]   President's Commission for the Study of Ethical Problems in Medicine and Biomedical and Behavioral Research. 1981. *Defining Death: A Report on the Medical, Legal and Ethical Issues in the Determination of Death.* Washington, D.C.: Government Printing Office.

[12]   In uncomplicated cases, that time period could be as little as six hours when a confirmatory test such as an electroencephalogram (EEG) or indictor of cerebral blood flow had been conducted. In the case of an identified cause for irreversible coma, 12 hours would be deemed necessary in the absence of a confirmatory test, and in the presence of anoxic (lack of oxygen) brain damage, 24 hours of observation was recommended. In the absence of confounding factors, such as hypothermia, shock, or drug intoxication, where an EEG pattern is isoelectric (flat EEG pattern) or cerebral blood flow is demonstrated to be absent, ten minutes of observation is deemed satisfactory.

[13]   The Quality Standards Subcommittee of the American Academy of Neurology. 1995. "Practice parameters for determining brain death in adults" (summary statement). *Neurology* 45, 1012–1014.

[14]   See footnote 4, p. 17.

There are neurological conditions that are capable of mimicking brain death, which make it important that the cause of the coma is determined. These conditions, such as hypothermia, drug intoxication, metabolic disturbances,[15] and some cases of Guillain-Barré syndrome, can be responsible for misdiagnoses of brain death. In the case of Guillain-Barré, the patient may look dead because of progressive involvement over a period of days of the peripheral and cranial nerves, but the syndrome is usually completely reversible. Similarly, hypothermia can present a clinical picture where brain-stem reflexes are absent. But even when the hypothermia is extreme, the brain deficits are potentially reversible.[16] Where there is suspicion of drug intoxication, such as from lithium, fentanyl, barbiturates, or tricylic antidepressants, to name a few, the patient should be observed for up to 48 hours to allow sufficient time for brain-stem reflexes to recover.[17] After the period of waiting, a confirmatory test is indicated if there is no clinical change in the patient.

The primary confirmatory tests for brain death include electro-encephalography (EEG), which records the electrical activity of the cortex and has been the most used of these tests; assessment of cerebral blood flow from tests such as transcranial Doppler sonography, cerebral angiogram, and nuclear medicine studies, especially SPECT and PET; and determinations of brain structure and function with magnetic resonance imaging (MRI) and brain-stem auditory evoked potentials (BAEP).[18] Each of these technologies provides information about the brain,[19] but none of them are without reliability

---

[15] See footnote 6, p. 407.

[16] E. F. M. Wijdicks. 1995. "Determining brain death in adults." *Neurology* 45, 1003–1011.

[17] If the drug is known, the recommendation is that the patient should be observed for up to four times the elimination half-life of the substance (this assumes there is no impediment to the elimination of the drug, such as the presence of another drug that interferes with its metabolism or other organ failure). For a complete discussion of this, see Wijdicks, footnote 5, p. 1218.

[18] A. M. Flannery. 1999. "Brain death and evoked potentials in pediatric patients." *Critical Care Medicine* 27, 264–265.

[19] The type of information is based on the object of study, be it electrical wave patterns, cerebral circulation, brain structure, or even brain function, such as can be seen with glucose metabolism.

problems.[20] These problems range from persistence of abnormal tracings after clinical death to inaccuracies in interpretation of markers, such as the reading of a scan image.[21] It is for these reasons that clinical neurological examinations repeated over a period of time are superior to technological tests. When conducted properly, clinical evaluations are very reliable, with excellent agreement among clinicians about the same patient and infrequent interpretation errors. The technological tests and guidelines are at best confirmatory of the clinical assessment.

---

[20]   An isoelectric (flat EEG pattern) recording does not in itself mean the person is brain dead. There have been reports of normal EEG electrical activity resuming after a period of a flat tracing (see E. F. M. Wijdicks. 2001. The Landmark "le Coma Depasse," in E. F. M. Wijdicks, ed., *Brain Death*. Philadelphia: Lippincott, Williams and Wilkins, 79). In addition, abnormal EEG tracings may persist for hours, and even longer, after clinical death has been established (see footnote 18, at 265). Transcranial Doppler sonography, which can be done at the bedside with a portable device, assesses flow in the middle cerebral arteries and the vertebral arteries. When the brain is dead, the reverberating flow caused by contraction of the arteries is absent. On the other hand, the signal may be absent if the skull interferes with transmission. Cerebral arteriography can give positive results in misleading conditions, and it can have false-negative results, particularly where open-head trauma is involved (see footnote 6, at 409). Radioisotope techniques (SPECT and PET) also show weaknesses. There may be inaccuracies in the tracer injection, and the scans must be read and interpreted by a specialist. In roughly 96 percent of those brain dead, the cerebral circulation is arrested. However, the brain stem may continue to be perfused, especially in children. Magnetic resonance imaging will often show tonsillar herniation, lack of flow intracranially, marked contrast enhancement of the nose, and poor differentiation of gray and white matter. This technique, along with MR angiography needs to be studied more extensively. Brain-stem auditory evoked potentials (BAEP) indicate brain-stem functioning (particularly the pons, but not the tracts of the medulla oblongata), which may be present in serious head trauma and barbiturate use even though the EEG may be isoelectric. Some patients have normal BAEPs although they are in a persistent vegetative state, and patients have had absent waves II to V but the presence of brain-stem reflexes following severe head injury, or anoxic-ischemic conditions. See E. F. M. Wijdicks. 2001. "Clinical Diagnosis and Confirmatory Testing of Brain Death in Adults," in E. F. M. Wijdicks, ed., *Brain Death*. Philadelphia: Lippincott, Williams and Wilkins, 83.

[21]   Establishing brain death in the pediatric patient has some differences. Though in both the adult and the child coma and apnea must be present, pediatric guidelines have differences in length of observation desirable and in what additional tests enhance the diagnosis (see footnote 18, at 265).

RESEARCH DIRECTIONS FOR BRAIN DEATH

In the short term, several existing technologies and combinations thereof may be shown to be superior to clinical evaluation. MRI and MRI angiography have not been thoroughly studied. Conventional angiography has been shown to be very accurate in determining brain death because brain function cannot exist if there is no blood flow shown in the intracranial blood vessels. The problem with conventional angiography is that it is an invasive procedure with potentially serious complications, such as causing a cessation of blood flow, which can affect viable organs that may be used in transplantation. Also, if high pressure is used or the body is in a position that causes the head to hang, some contrast material may be introduced into the intracranial vessels, which would give the impression that blood flow was continuing, when in fact it had ceased.[22] MRI arteriography is increasingly being shown to be an accurate, reliable technique to establish absence of flow in the intracranial vessels. In addition, it is a noninvasive procedure. MRI has undergone some major improvements in hardware and software over the past several years, which have enhanced the effectiveness of that technology. Some of these include faster sequences, dedicated coils, and stronger gradient fields. Studies have shown that specific MRI signs exist for brain death, signs like tonsillar herniation, absence of circulation in intracranial vessels, and lack of visualization of the intracranial veins. The MRI signs are highly sensitive, making them a potentially superior confirmatory test when clinical findings are documented.[23] With increasing improvements, MRI and MRI angiography may be demonstrated to be superior to clinical evaluation.

Some early studies of multimodality evoked potentials (MEPs) are showing many advantages over current confirmatory tests. MEPs include BAEPs, flash-visual evoked potentials (flash VEPs),

---

[22] A. H. Karantanas, G. M. Hadjigeorglou, K. Paterakis et al. 2002. "Contribution of MRI and MR angiography in early diagnosis of brain death." *Eur Radiol* 12, 2710–2716, 2715.

[23] Ibid, 2713–2716.

and median somatosensory evoked potentials (median SEPs).[24] The combination of these three evoked potentials constitutes MEPs, which test the cerebral cortex as well as the brain stem. MEPs are insensitive to misleading factors like drugs, metabolic disturbances, and hypothermia and can be performed easily at the patient's bedside. With refinement of the techniques and coordination of the three tests, it is possible that MEPs, as in the case of MRI and MRI angiography, may prove to be highly reliable for determining brain death.

The long term would depend on further developments and refinements of imaging technologies like PET and MRI. When PET is used for determining glucose metabolism, it offers an accurate determination of brain death. If the images are essentially black and no metabolism is occurring, one can reliably assume the brain or parts of the brain are dead. The future should bring a new generation of imaging technologies based on MRI, PET, and SPECT that are capable of greater sensitivity and specificity. These technologies will likely be smaller, portable, and capable of detecting small amounts of substances such as specific protein concentrations in the brain that increase during periods preceding and at brain death.[25] They would demonstrate not only the presence of proteins associated with nerve death (apoptosis), but also the deprivation of neurotrophic factors necessary for the continued life of the neurons.[26]

New developments, such as magnetoencephalography (MEG)—also referred to as superconducting quantum interface device, or SQUID—are already enhancing the power of imaging. MEG is capable of picking

---

[24]   BAEPs involve auditory clicks that affect functioning of the cochlear nerve and the pontine auditory pathways. Flash VEPs are induced by light and involve following retinal activation to the visual occipital projection and the associative areas. And median SEPs involve electrical stimulation of the median nerve at the wrist. This is followed by recordings peripherally, at the brachial plexus, cervical level, low brain stem, and double cortical level—parietal and frontal (see footnote 6, at 410–411).

[25]   Studies in mammals are revealing the presence of genes that program cell death. They produce proteins for apoptosis regulation—the Bcl-2 family of proteins, the adaptor protein Apaf-1 (apoptotic protease-activating factor 1), and the cysteine protease caspase family have been identified thus far. With greater delineation of these proteins and their genes, one or more may be found to be reliable markers for dying and dead neurons (see footnote 26, at 802).

[26]   J. Yuan and B. A. Yanker. 2000. "Apoptosis in the nervous system." *Nature* 407, 802–809.

up and measuring extremely weak magnetic fields that are induced by electric activity from firing neurons. Although MEG can get below the cortical surface and map the whole brain, its best results come from the cortical surface area. MEG allows one to see activity of small clusters of cells. The future will see improvements in the power of this technique, enabling it to detect cellular activity in the brain stem. Arguments are strong for redefining death on higher-brain criteria based on cortical nonfunctioning. What is important is not simply the presence of activity itself in the brain, but rather the presence of organized activity. Hence, evoked potentials challenge the system and allow for observation of levels of organization. When organization is nonexistent or hampered irreversibly, it is definitional for brain death.

Furthermore, what constitutes human identity is believed to involve primarily the frontal lobe and its connections with the temporal and parietal lobes.[27] When the frontal lobe is irreparably damaged and disorganized, the patient may continue for a long time in a vegetative state. However, without higher-level cerebral functioning, the patient is arguably no longer a person.[28]

A FUTURE DIRECTION FOR BRAIN DEATH

Some of the recent work of researchers like Miguel A. L. Nicolelis at Duke University has introduced other speculative possibilities in the long term for dealing with brain death. Dr. Nicolelis has demonstrated that a brain-machine interface can be established. Using electrodes on the motor cortex, he has shown that it is possible with the aid of sophisticated

---

[27] There are many sites that address the cognitive and emothional understandings of self. See V. S. Ramachandran and S. Blakeslee. 1998. *Phantoms in the Brain: Probing the Mysteries of the Human Mind.* New York: William Morrow, 225–257; E. Goldberg 2001. *The Executive Brain: Frontal Lobes and the Civilized Mind.* Oxford: Oxford University Press, 139–156; S. R. Quartz and T. J. Sejnowski. 2002. *Liars, Lovers, and Heroes: What the New Brain Science Reveals About How We Become Who We Are.* New York: William Morrow, 138–141; and A. Newberg, E. D'Aquili, and V. Rause. 2001. *Why God Won't Go Away: Brain Science and the Biology of Belief.* New York: Ballantine Books, 18–34.

[28] J. Fisher. 1999. "Re-examining death: Against a higher-brain criterion." *Journal of Medial Ethics* 25, 473–476.

mathematical models to translate motor control commands from raw electrical brain activity.[29] These commands are transferred to a multijoint robot arm that moves on command.[30] If it is indeed possible to transfer impulses for motor action to a machine, would it not be likely at some future point to transfer through complex mathematical models thoughts, ideas, and feelings from the cortex of a person to a computer or a robot?[31] Hence, while a patient is in the hospital and it is determined that neurons are dying (the process of apoptosis), the patient's "mind" might be downloaded into a computer, perhaps even transferred to a robot. It may indeed be possible with current technology to hook up electrodes and determine whether there is any organized activity at all. If it is impossible to download anything where downloading is feasible, then, in the absence of mechanical disruption, that would demonstrate that the brain is dead. This should serve as the new definition of brain death.[32]

Cognition and Legal Competency

Competency, or cognitive understanding of events and one's relationship to them, is relevant to a wide range of legal matters. In criminal law, assessing competency applies to the ability of an accused to stand trial. In civil law, competency applies to many transactions, such as one's ability to execute a will, enter into a contractual arrangement, manage one's affairs, and give informed consent for medical research or treatment. In a broad sense, competency also applies to witnesses, such as a bystander who witnesses an automobile accident but is cognitively impaired such that he or she cannot provide a coherent narrative about the event.

---

[29] M. A. L. Nicolelis and S. Ribeiro. 2002. "Multielectrode recordings: The next steps." *Current Opinion in Neurobiology* 12, 602–606.

[30] M.A.L. Nicolelis. 2003. "Brain-machine interfaces to restore motor function and probe neural circuits." *Nature Reviews/Neuroscience* 4, 417–422.

[31] See footnote 30, at 417 and 418, on the use of mathematical models to extract brain activity.

[32] For an interesting discussion of the future possibilities, see J. Waldbauer, in collaboration with M. Gazzaniga. 2002. "The Self-Made Brain, Neuroscience in 2025." In Proceedings of *Neuroethics: Mapping the Field*. Conference held in San Francisco, May 13–14.

In this sense of witness competency, the witness isn't precluded from testifying, but evidence of the cognitive impairment could affect the weight given his or her testimony. For the most part, the law assumes that functioning adults have the competency to fulfill these many functions. When competency is called into question—as in cases of serious head injury,[33] mental illness with manifested delusions, and neurodegenerative disease, such as would be surmised in an elderly citizen whose memory seems less sharp than previously or whose family members question the terms of his or her will—then courts may request a psychiatric examination, with neuropsychological testing where indicated.

Clinical mental health evaluations, though limited in scope, have been the only tests available to assist judges in determining the status of a person's competency. The structured psychiatric interview, which focuses on determining whether a person suffers from a mental disorder, such as an organic mental illness, schizophrenia, bipolar illness, drug abuse, and so on, is largely phenomenological or descriptive in character. The evaluator asks questions about the person's history and behavior and conducts a mental status examination. If there is a serious cognitive disorder or a neurological deficit, the evaluator might order tests, including X rays and an MRI, or refer the person to a neurologist for consultation. The evaluator will also probably refer the person for neuropsychological testing, which includes tests for intelligence, memory, attention, and personality disorders.[34]

Developments in the neurosciences over the past ten years have introduced other dimensions that are relevant to the assessment of competency. Much attention has been devoted to the frontal lobe, particularly the prefrontal lobe, as having the major role in cognition. This area of the brain obtains information from all major systems—the forebrain, limbic system, and so on—synthesizes the information,

---

[33] See D. M. Bitz and J. S. Bitz. 1999. "Comment: Incompetence in the brain-injured individual." *St. Thomas Law Rev.* 12, 205.

[34] Often included are tests of intelligence (such as the Wechsler Adult Intelligence Scale, or WAIS III), personality inventories (such as the MMPI, or Minnesota Multiphasic Personality Inventory), tests of memory (Wechsler Memory Scale-Revised, or WMS-R, Neimark Memorization Strategies), and tests of attention (Wisconsin Card Sorting Task, or WCST), to name a few.

and sets the guiding principles obtained from experience to structure intelligent, goal-directed behavior.[35] But research is hitting all areas of human cognition,[36] from brain processes associated with decision making and the complexity of memory to the impact of fear and other emotions on "rational" and "moral" decision making. Research findings have considerably expanded the framework for assessing competency, as distortions in perception may occur at various stages of cognition that would not have been picked up by the more traditional assessments. In the following discussion, I address some of the productive areas of research as they apply to civil competency matters—specifically cognition—and speculate on the short-term and long-term neuroscience findings that might influence our understanding of competency and the way it is likely to be evaluated for relevancy to the law.[37]

THE NEUROSCIENCE OF MEMORY

One of the most important areas of cognition that has come under close scrutiny is memory, and the hippocampus and other adjacent structures in the medial temporal lobe have been the primary focus of this research. The research has shown the pivotal role of the hippocampal region[38] in declarative memory,[39] which is the ability to remember facts and events from the past.[40] The role of the hippocampus in declarative memory

---

[35]   E. K. Miller, D. J. Freedman, and J. D. Wallis. 2002. "The prefrontal cortex: Categories, concepts and cognition." *Philosophical Transactions of the Royal Society of London—series B: Biological Sciences* 357, 1123–1136.

[36]   A. B. Hariri, V. S. Mattay, A. Tessitore et al. 2002. "Serotonin-transporter genetic variation and the response of the human amygdala." *Science* 297, 400.

[37]   For a discussion on the brain and scientific evidence particularly as it applies in criminal law see E. Beecher-Monas and E. Garcia-Rill. 1999. "The law and the brain: Judging scientific evidence of intent." *J. App. Prac. & Process* 1, 243.

[38]   Refers primarily to the hippocampus and parts of the brain that are adjacent, such as the dentate gyrus and subicular complex, and adjacent perirhinal, entorhinal, and parahippocampal cortices.

[39]   Nondeclarative memory refers to habits, skills, conditioning—these express the impact of experiences through performance.

[40]   J. R. Manns, R. O. Hopkins, J. M. Reed et al. 2003. "Recognition memory and the human hippocampus." *Neuron* 37, 171–180.

has been shown to be extensive,[41] and is not limited to remembering complex tasks, such as forming relationships and connections among various stimuli. It includes semantic memory, the capacity for learning facts and remembering them.[42] It also includes both features of recognition memory—that is, a "familiarity" component, which is judging recently obtained facts or items as familiar, and an "episodic" component, which is remembering episodes in which facts or items were learned. The latter includes onetime novel experiences.[43] Recent research has also shown that the hippocampus and medial temporal lobe (MTL) are engaged by representations that are essential for the individual to learn a task—for example, to learn motor sequences even when implicit (that is, the individual is not informed about the underlying sequence).[44]

Two research methods have emerged in recent years that are facilitating our understanding of how the brain processes information. The first of these is the tracking of cognitive processes in the brain. This is referred to as mental chronometry, which essentially sets out cognitive, motor, or perceptual tasks into sequences, or stages of processing, based on response times.[45] Brain areas often work in parallel in processing information. However, many cognitive tasks are complex and are characterized by serial stages of processing—that is, intermediate outputs are necessary for the next processing stages. In humans, two modalities have been most useful in relating neural events and their sequences to cognitive stages that are sequenced toward the execution of a task: MEG, which detects activity in small clusters of cells on the surface of the brain, and EEG, which picks up changes in neuronal

[41] Ibid., 171.

[42] J. R. Manns, R. O. Hopkins, and L. R. Squire. 2003. "Semantic memory and the human hippocampus." *Neuron* 38, 127–133.

[43] K. Nakazawa, L. D. Sun, M. D. Quirk et al. 2003. "Hippocampal CA3 NMDA receptors are crucial for memory acquisition of one-time experience." *Neuron* 38, 305–315.

[44] See R. A. Poldrack and P. Rodriguez. 2003. "Sequence learning: What's the hippocampus to do?" *Neuron* 37, 892–893.

[45] J. D. Schall. 2003. "Neural correlates of decision processes: Neural and mental chronometry." *Current Opinion in Neurobiology* 13: 182–186.

activity on a scale of milliseconds.[46] However, neither is able to obtain an adequate three-dimensional spatial design of neural activity. In contrast, fMRI[47] results in excellent spatial resolution but basically inadequate temporal resolution.[48]

Many functional domains have been studied with time-resolved fMRI. For example, tracking has been done of sequences of activation from language input regions (Wernicke's area and the auditory cortex) to sections of the brain involved in articulation (Broca's areas) and motor output regions.[49] Also, tracking is being done to understand the process of making decisions. Such studies for the most part are focused on processes that change sensory discrimination into an action, even though the data point strongly to prior experience as the frequent guide to behavioral choices. Nonetheless, tracking follows the sensory input from the sensory cortex to the parietal and prefrontal association cortices, where evidence is evaluated and prospects conceptualized, to the cortical and subcortical motor structures, where categorical behavior output occurs, and ultimately to the basal ganglia, orbitofrontal cortex, and cingulate cortex,[50] where outcome evaluation occurs.[51] Some researchers have begun to study the effects of prior experience on choice of behavior as well as the associated decision processes from sensory to motor regions.[52] The area of abstract, categorical representations has also

---

46 E. Formisano and R. Goebel. 2003. "Tracking cognitive processes with functional MRI mental chronometry." *Current Opinion in Neurobiology* 13: 174–181.

47 Particularly blood-oxygenation-level-dependent (BOLD) fMRI; see footnote 40, p. 174.

48 See footnote 40, p. 175.

49 See footnote 40, p. 177.

50 Studies are showing that the anterior cingulate cortex functions central to intelligent behavior, that is, focused problem solving, adaptive response to changing conditions, emotional self control, error recognition. It contains a distinctive class of large spindle-shaped neurons, which emerge postnatally. Also, the anterior cingulate cortex receives projection from the amygdala, and dopaminergic projections, which are believed to be involved with reward. See J. M. Allman, A. Hakeem, J. M. Erwin et al. 2001. "The anterior cingulate cortex: The evolution of an interface between emotion and cognition." *Annals of the New York Academy of Science* 19, 107–117.

51 M. L. Platt. 2003. "Neural correlates of decisions." *Current Opinion in Neurobiology* 13, 141–148.

52 Ibid., 143.

been the subject of study. Humans can readily adjust to novel events. This strongly implies the presence of internal representations that are abstracted and generalized, making them adaptable to new situations. Several brain regions are involved as neural correlates, most particularly the inferior temporal cortex (ITC) and the prefrontal cortex (PFC).[53]

Mental chronometry studies offer an excellent opportunity to plot out the staging of processes in the brain. These studies do more than just determine brain regions that are activated during cognition. They allow for the integration of thinking with brain activity and ultimately behavior, and thereby reveal the relative contributions and sequencing of activated regions of the brain to stages of processes that lead to the actual performance of a task.

Genetics and the Brain

Another method that is having success in furthering understanding of the brain involves the relationship of genetics to neuropsychiatry and behavior. This multilevel approach uses imaging and other techniques to delineate the neuroanatomical structure and function of the brain and correlates this information with genetic factors and environmental influences. It has been successful in elucidating the full picture of what is happening with syndromes in which cognitive and personality deficits are manifested.[54] Recent studies correlating genetics and images of the brain to behaviors have been pointing the way to specific gene differences affecting circumscribed behaviors. A study was reported last year that located a single gene that influences a person's response to emotionally charged stimuli.[55] This specific gene

---

[53] For a discussion on categories and concepts and their neural correlates, see E. K. Miller, A. Nieder, F. J. Freedman, and J. D. Wallis. 2003. "Neural correlates of categories and concepts." *Current Opinion in Neurobiology* 13, 198–203.

[54] In its broadest sense, this multilevel approach involves relating neuroanatomical structure and brain function to known genetic conditions or syndromes, such as Williams syndrome, fragile X syndrome, and others that are known to have neurobehavioral components. The term "behavioral neurogenetics" has been applied to this research approach. See A. L. Reiss, S. Eliez, J. E. Schmitt et al. 2000. "Brain imaging in neurogenetic conditions: Realizing the potential of behavioral neurogenetics research." *Mental Retardation and Developmental Disabilities Research Reviews* 6, 186–197.

[55] See footnote 36.

encodes the production of a protein whose function is to transport serotonin back into neurons following its release at synapses, thereby minimizing the amount of time the serotonin can affect adjacent neurons. This gene consists of two structural alleles, one with a shorter promoter region than the other. The gene with the short allele produces less of the protein used for transport than does the longer allele. Given this information, the researchers examined the amygdala, which is believed to be the center for emotional commands and, therefore, conditioned fear responses. Using fMRI focused on the amygdalae of subjects, the researchers showed that the provocative effects of pictures of scary faces was greater on those with short alleles. Scanning demonstrated that those with the short allele showed greater activation of their amygdalae and a greater fear response.

This study linked genetic variation to differences in emotional responses and in brain activity.[56] Fear and concomitant anxiety affect an individual's capacity to respond to his or her environment and, therefore, prevent effective cognition.[57] Recent studies are showing the impact of amygdalar response on novelty, which may have a genetic basis. One study compared children assessed by the age of two who had inhibited temperaments, manifested by withdrawing from novelty, such as unfamiliar situations, people, and objects, with children of the same age who had uninhibited temperaments, manifested by a spontaneous approach to novel and unfamiliar situations, people, and objects.[58] The researchers found that these temperaments are relatively stable through adolescence and that adults who were categorized as uninhibited during infancy

---

[56] Another genetic study on behavior was reported on the relationship of an enzyme (monamine oxidase A, or MAO-A) and violent behavior. This study showed that when the gene that codes for the enzyme is defective, neurotransmitters are not effectively metabolized, which creates the setting for the potential emergence of a violent adult. The critical factor becomes whether the individual with the defect has been abused as a child. Here again one is confronted with the impact of something genetic on the character of the individual's decision making. See A. Caspi, J. McClay, T. E. Moffitt et al. 2002. "Role of genotype in the cycle of violence in maltreated children." *Science* 297, 851–854.

[57] See R. B. Adams, H. L. Gordon, A. B. Abigail et al. 2003. "Effects of gaze on amygdala sensitivity to anger and fear faces." *Science* 300, 1536. Contains discussion on impact of anger and fear on amygdala function.

[58] C. E. Schwartz, C. I. Wright, L. M. Shin et al. 2003. "Inhibited and uninhibited infants 'grown up': Adult amygdalar response to novelty." *Science* 300, 1952–1953.

when studied with fMRI demonstrated greater response to novelty. Even though the "uninhibited" child may demonstrate impulsivity, aggression, and possibly antisocial characteristics during adolescence, an upside may be that these individuals will be open to new experiences and ideas, thus having a different approach to the outside world.

EMOTIONS AND COGNITION

Research has been supporting the notion that emotions play a major role in decision making, even rational decisions. Antonio Damasio has proposed the "somatic-marker hypothesis," which he derived from studies of adult patients who suffered injury to sections of their prefrontal region, most particularly the ventromedial frontal lobe, especially the orbitofrontal cortex.[59] He found that these individuals are adversely affected in their capacity to make personal and social rational decisions. Emotional factors shaped by similar situations in the past act as "somatic markers," "informative feelings," that guide us in future decision making. These capacities arise from complex interactions between the limbic system, especially the hippocampus and amygdala, which contains memories of the context in which emotions are experienced, and the prefrontal lobe. Studies have extended this observation to moral decision making. One group of researchers evaluated adults who experienced prefrontal cortical injury, especially of the orbitofrontal cortex, before 16 months of age and found that unlike those injured as adults, they demonstrated severe defects in their social and moral reasoning.[60] Subsequently, researchers have shown through fMRI that both emotion and reason have important functions in the shaping of personal moral judgment.[61] The patterns of activation and psychological processing for impersonal moral judgments, consistent with Damasio's somatic-marker

---

[59] See A. R. Damasio. 1999. *The Feeling of What Happens: Body and Emotion in the Making of Consciousness.* New York: Harcourt.

[60] S. W. Anderson, S. Bechara, H. Damasio et al. 1999. "Impairment of social and moral behavior related to early damage in human prefrontal cortex." *Nature Neuroscience* 2, 1032–1037.

[61] J. D. Greene, R. B. Sommerville, L. E. Nystrom et al. 2001. "An fMRI investigation of emotional engagement in moral judgment." *Science* 293, 2105–2108.

thesis, were different, resembling more closely the patterns of decision making that are not involved with a moral dilemma.[62]

MEMORY AND COMPETENCY

The complexity of memory will be more clearly understood in the near future. Already much has been learned about the structures in the brain, especially the hippocampal region, most involved with declarative and, more specifically, semantic and episodic memory. In the near future, it should be possible to determine if those centers of the brain are activated and properly sequencing with the prefrontal lobe when a person is asked questions about a transaction, thereby indicating that person's level of competency. It will also shortly be possible to identify pathways of sensory input to specific memory functions, the way they are processed, and the full complement of neural networks involved. In addition, the effects of medications on memory will be considerably improved. Attention has been focused recently on the influence of antidepressants, particularly selective serotonin reuptake inhibitors such as Prozac, on memory. A study of women with recurrent depression showed that the volume of their hippocampus was 10 percent less than that of the control group, who were not depressed. Antidepressants actually seem to prevent neural damage and enhance neurogenesis.[63] A recent study on mice not only supports this position about the effects of antidepressants on neurogenesis in the hippocampus but also suggests that the cause and effect relationship of these medications with neuron growth may be the way they bring about behavioral benefits.[64] Similarly, genes associated with enhancing or inhibiting those chemicals found to be necessary for memory will be mapped out, as well as ways to alter their effects.[65]

---

[62] Ibid.; footnote 61, p. 2107.

[63] Y. Sheline. 2003. "Untreated depression and hippocampal volume loss." *American Journal of Psychiatry* 160, 1516–18. See also C. D'Sa and R. S. Duman. 2002. "Antidepressants and neuroplasticity." *Bipolar Disorders* 4, 183–194.

[64] L. Santarelli, M. Saxe, C. Gross et al. 2003. "Requirement of hippocampal neurogenesis for the behavioral effects of antidepressants." *Science* 301, 805–809. See also G. Vogel. 2003. "Depression drugs' powers may rest on new neurons." *Science* 301, 757.

[65] In recent research it has been shown that catechol O-methyltransferase (COMT) is implicated in inactivating catecholamines, including dopamine, thereby affecting

Research Directions and Cognition

In the short term, research will delineate more neural circuits involving cognition. Through techniques like chronometry and imaging, the pathways from sensory input—visual, auditory, and tactile—as well as from memory[66] to the limbic system, parietal lobe (which is involved in attention and behavioral relevancy),[67] and frontal lobes[68] will be worked out. Assisting in the mapping of the brain are technologies such as transcranial magnetic stimulation (TMS), which stimulates the brain from outside the skull, with minimal discomfort for the subject. TMS pulses cause disruption in target regions of the brain for varying periods, depending on the rapidity of the repeated pulses. If the rate of pulses is slow, TMS will suppress the brain function affected. If, on the other hand, the rate of pulses is rapid, then affected areas of the brain will be excited and enhanced. In the case of suppression of the brain, the disruption TMS causes creates a "virtual" patient, one who appears as though he or she has suffered actual damage to the brain.[69] This is a safe, effective way to explore the function of regions of the brain.[70]

---

working memory involving frontal lobe function. The functional polymorphism of the gene, which affects memory, has been identified and COMT located in the human cortex. *See* M. Matsumoto, C. S. Weickert, M. Akil et al. 2003. "Catechol O-methyltransferase mRNA expression in human and rat brain: Evidence for a role in cortical neuronal function." Neuroscience 116, 127–137.

66 See M. Ohbayashi, K. Ohki, and Y. Miyashita. 2003. "Conversion of Working Memory to Motor Sequence in the Monkey Premotor Cortex." *Science* 301, 233–236.

67 J. A. Assad. 2003. "Neural coding of behavioral relevance in parietal cortex." *Current Opinion in Neurobiology* 13, 194–197.

68 Also interesting was research recently reported on the linkage between action and outcome. The researchers studied the neural mechanism for goal-based action selection, finding activation in the medial prefrontal cortex. See K. Matsumoto, W. Suzuki, and K. Tanaka. 2003. "Neuronal correlates of goal-based motor selection in the prefrontal cortex." *Science* 301, 229–232.

69 A. Cowley and V. Walsh. 2001. "Tickling the brain: Studying visual sensation, perception and cognition by transcranial magnetic stimulation." *Progress in Brain Research* 131: 411–25.

70 Many studies have been done with TMS (particularly through virtual lesions) to influence brain functions to explore how the brain works. Studies have investigated attention, speech, mood, memory, movement, and visual perception, to name a few. See M. S. George, Z. Nahas, F. A. Kozel et al. 2003. "Mechanisms and current state of transcranial magnetic stimulation." *CNS Spectrum* 8, 496–513.

Studies have been conducted using PET to examine intelligence. One such study reported in the early nineties involved PET images of the glucose metabolic rate of subjects while they were performing a test of abstract reasoning.[71] The test, Raven's Advanced Progressive Matrices, which consists of 36 problems, correlates very closely with IQ. The glucose metabolic rate assessed from the images of the participants had an inverse relationship to high scores. In subsequent studies of Down's syndrome, researchers also found an inverse relationship between glucose metabolic rate and IQ. This suggested to the researchers that efficiency of brain energy manifested by a low metabolic rate was related to good cognitive performance. The relationship of images of various anatomical regions of the brain associated with cognition to intelligence tests suggests that in the near future images will be carefully calibrated to become specific markers for intelligence. PET studies with different radio-labeled drugs to manipulate states of the brain while tasks are being conducted will likely lead to an understanding of which neurotransmitters, and their sequencing, are related to specific cognitive performance. Hence, specific images will reflect not only levels of intelligence but relationships among subsets of cognitive functioning.[72]

In addition to work on intelligence, centers of emotional responses, such as disgust and empathy,[73] will be located and their effects on cognition identified.[74] These emotions, along with fear, create feeling tones that alter the clarity of cognition. In the future, it should be possible to determine the impact of emotions on any one decision and to develop a method for weighing when the emotions

---

[71] R. J. Heir. 1993. "Cerebral glucose metabolism and intelligence." In P. A. Vernon, ed., *Biological Approaches to the Study of Human Intelligence*. Norwood, NJ: Ablex Publishing.

[72] See R. J. Heir. 1998. "Brain Scanning/Neuroimaging." In *Encyclopedia of Mental Health, Vol. 1*. New York: Academic Press.

[73] Some studies have already suggested that the emotion of disgust is located in subregions of the ventral and anterior insula, and empathy appears to be in subcortical structures such as the amygdala and hippocampus. See "The Neuropsychology of Disgust." *Neuropsychiatry Reviews*, March 3, 2003: 18–20.

[74] See R. Adolphs. 2003. "Neural systems for recognizing emotion." *Current Opinion in Neurobiology* 13, 169–177.

trump an individual's ability to make a personal rational choice. The "somatic marker" hypothesis—that "informative feelings" shaped by experience affect decision making and judgment—introduces yet another feature of emotions and their impact on cognition. In the future, imaging technologies should be able to determine whether informative feelings are operating in decisions.

The influence of emotions on cognition has been considered in criminal law since the mid-1950s. In the Durham test, " . . . product of mental disease or mental defect," Judge David Bazelon considered the impact of feelings, particularly a brooding mood state, on "competency" to commit a crime.[75] Subsequently, the American Law Institute's test of insanity incorporated emotions with terms like "appreciate the wrongfulness of conduct" and "conform conduct to the requirements of the law."[76] For the most part, these tests, except for the concept "appreciate," recognize the power of emotions to influence an individual's "control" of behavior. The term "appreciate" is broader and suggests the ability to inculcate feelings with thoughts. In that sense, it approaches the notion of "informative feelings." Both of these tests for insanity were criticized because of the practical difficulty of determining in any one case the weight of emotional factors in converting thoughts to action. Now, with tracking and research findings regarding activation of parts of the limbic system and frontal lobe under selective conditions, it will be possible to assess an individual in broader terms than just his or her capacity for factual understanding as determined by neuropsychological tests. It will be feasible to visualize whether he or she emotionally understands concepts, which will be established by the level of functioning in the brain and the appropriateness of choices.

Tracking will also provide information on mental interference with an individual's thinking. Interferences such as would occur because of auditory or visual hallucinations, delusions, and disorganized thinking

---

[75] See D. L. Bazelon. 1974. "Psychiatrists and the adversary process." *Scientific American* 230, 18. See also *Durham v. United States* 214 F. 2d 862 (D.C. Cir., 1969).

[76] ALI Model Penal Code. 1955. Section 4.01 (1) (Tent Draft no. 4).

will be determined by a scan. Up to now, psychiatry has had to rely on a patient's admission of hallucinations, or inference of such based on specific behavior. Soon it will be possible to verify through tracking whether an auditory or visual hallucination is occurring and what its effect is on competency. Similarly, delusions, often seen clinically with disturbances in receptors associated with obsessive thinking, will be relegated to images that can be used to verify the competency of a person who appears to be experiencing them. Thought process disorders will be delineated by a comparison of images with normal sequencing in the parietal, medial temporal, and prefrontal lobes.

Finally, there will be more information on the relationship of genetic abnormalities to cognition and emotions. Studies comparable to that of the gene with short alleles that increases fear will be replicated with respect to a full range of emotions and mental aberrations. Delusions may also be found to have some association with specific gene dysfunction.

Future Directions in Cognition

In the long term, the following are likely developments. The relationship of genes to cognition and to emotional temperament will be completely mapped out. This will provide for various tools—from enzyme assays to images—for predicting people's intellectual abilities and emotional responsiveness. Profiling will occur early, possibly even at the time of birth, when cord bloods can be genetically detailed and brain images obtained from attenuated MRI. People could be categorized into types based on expected personality characteristics and talent. Scanners will become smaller, until a portable scanner will be available so that an individual can sit within a small space and be scanned in a resting state, while answering questions, and under external provocation. This will provide information on cognitive abilities and emotional input from the medial temporal lobe, as well as tolerance for stress. The outcome is a psychological portrait that delineates not only boundaries of intellectual abilities but limits of emotional capacity as well.

In the future the role of neural stem cells will be more thoroughly understood, particularly the ways they can be encouraged to develop into select brain cells.[77] These stem cells may be used to restore competency, ranging from intelligence to linguistic and mathematical abilities. Techniques will be discovered for accelerating the transformation of stem cells so that an incompetent person can be readily rendered capable of entering a wide range of transactions. In addition, brain transplantation—which has been successfully used in the case of embryonic dopamine-rich nerve cells injected into the striatum to treat Parkinson's disease, but not yet with sections of the brain[78]—will be further advanced, so that it is likely that transplantation of parts (including clusters of cells) or the whole brain will be possible. For example, an individual with memory loss resulting from injury or damage to his or her hippocampus may benefit from a transplanted hippocampus secured from a person unable to survive because of massive damage to other critical sections of the brain.[79] Augmenting these developments will be our understanding of the process of nerve apoptosis, which will include information on the nature and mechanism of the action of proteins that accelerate the process. Methods, such as blocking chemicals, will be available to reverse the process in a dying person so that he or she can be made competent, even if only on a temporary basis, to complete important personal and family transactions before dying.

---

[77] Work has already been done on methods for cell amplification and differentiation in vitro—that is, tools to explore cell survival, proliferation, and differentiation are being developed. See H. S. Keirstead. 2001. "Stem cell transplantation into the central nervous system and the control of differentiation." *Journal of Neuroscience Research* 63, 223–226. See also A. Bescovi, A. Gritti, G. Cossu et al. 2002. "Neural stem cells: Plasticity and their transdifferentiation potential." *Cells Tissues Organs* 17, 64–76.

[78] See L. Melton. 2000. "Neural transplantation: New cells for old brains." *The Lancet* 355, 2142–2143.

[79] This will introduce many interesting ethical and psychological issues about the identity of the person receiving the transplanted hippocampus, as the memory base will likely be that of the donor. Work is being done with embryonic hippocampal tissue (which doesn't yet contain much memory), which is likely to be a viable target for neural grafting. See D. Turner. 2003. "Clinical prospects for neural grafting therapy for hippocampal lesions and epilepsy." *Neurosurgery* 52, 632–644.

Though currently speculative, the process of downloading thoughts, ideas, mental processes, memories, and feelings may be available, so that a person's mind can be assessed more fully and objectively in terms of intelligence, general competency, emotional content, and creative potential.[80] Our understanding of brain plasticity will be advanced to complement the benefits of downloading. We will learn how to restore a person to normal functioning levels through various techniques so as to ensure the completion of important transactions or work projects. The return to normal, however brief, may come from stimulation of the brain through learning techniques, medications, or devices such as microtransistors that can be navigated to specific cognitive centers to stimulate neural response and development. It is already possible, through imaging, to see brain changes that are brought on by skilled learning.[81] Increasingly there is evidence of growth and formation of neurons in different parts of the brain even into adulthood. A recent study of taxi drivers in London demonstrated that the hippocampus, especially the right posterior hippocampus, where complex maps are stored and used, is larger in that group than in aged-matched males in the average population.[82] This shows the effects on the hippocampus of memory stimulation that occurs during adulthood, reflecting the needs of a taxi driver to have command of the vagaries of a city's streets.

Enhancing Cognitive Functions Beyond Normal

One of the outcomes of the past few decades of neuroscience research is the opportunity in the future to enhance man's capacities in

---

80   G. Brumfiel. 2002. "Futurists predict body swaps for planet hops." *Nature* 418, 359.

81   L. G. Ungerleider, J. Doyon, and A. Karni. 2002. "Imaging brain plasticity during motor skill learning." *Neurobiology of Learning Memory* 78, 553–564.

82   E. A. Mcguire, D. G. Gadian, I. S. Johnsrule et al. 2000. "Navigation related changes in the hippocampus of taxi drivers." *Proceedings of the National Academy of Sciences, USA* 97, 4398–4403.

cognitive areas to levels never before attained. Providing the possibility for higher IQs, greater memory powers, acute sensitivity to the nuances of human communication, and heightened awareness of one's moral responsibility would be hardly viewed as undesirable outcomes of research. We would all welcome such enhancements. However, they would come with some potential for serious risks. Enhancements such as medications and gene therapy may have serious adverse reactions associated with them. Perhaps one's IQ or memory would be improved, but there may be risks and trade-offs to be considered.[83] Intelligence may be enhanced, but perhaps at the expense of developing an undesirable personality trait. We've seen since the nineties that antidepressants such as Prozac may profoundly affect an individual's feelings and sense of self. The effects of potentially permanent enhancement techniques—medications, TMS, learning strategies, and others—are likely to be more profound. The recipient may feel different, like a changed person. The question is whether we would be willing to allow alterations to parts or the bulk of our personality for such enhancement.

Some of the scientific discoveries allowing for enhancement strategies were reviewed in the section Cognition and Legal Competency. In the short term we will see more effective medications with greater specificity to enhance memory and cognition. Since the full genome sequence for humans has been explicated, in some cases it is now possible to link specific genes to brain function and dysfunction. The resulting data will increase geometrically our knowledge about the underlying biology of cognition and personality development. New diagnostic tools will be developed that will determine drug responsiveness in any one person, thereby enhancing not only treatment of those suffering from dysfunction, but opportunity for those desiring enhancement of their capacities. This will lead to sophisticated cognitive profiling, whereby tools for assessing the genetic makeup of an individual will be developed to determine what genes are modulated

---

[83] See A. Roskies. 2002. "Neuroethics for the new millennium." *Neuron* 35, 21–23.

by neural activity,[84] as these genes offer excellent opportunities for inducing brain plasticity.[85] Each of these genes with potential for plasticity will be important in developing drugs that can be targeted to bring about modifications to enhance cognition.

ENHANCING MEMORY AND COGNITION

The work of Eric Kandel, a Nobel Prize recipient, has been key in the development of memory drugs. Kandel showed that a messenger molecule, cyclic AMP, is important for forming memories. This molecule activates C-AMP response element binding protein (CREB), which stimulates proteins that promote a synaptic growth process that in turn fortifies the connections between neurons. CREB has been shown to be necessary for the conversion of newly established memory—that is, short-term memory—to long-term memory.[86] Work is under way to develop drugs to boost C-AMP as well as CREB levels. Other efforts are underway to develop a memory amplifier. A family of drugs known as ampakines, which focus on the neurotransmitter glutamate[87] and its

---

[84] Much of the research leading to the understanding of these genes came from experiments with flies, but the genetics is such that there are mammalian homologs (the mammalian genome is about five times larger than that of the fly), and particularly human ones that can be identified through informatics or cloning. These genes involved with brain plasticity allow for a "dependent mechanism" of fine-tuning the basic wiring of the brain so as to create adaptive behavioral responses to the environment. The theory is that neural activity affects gene expression, which subsequently changes the structure and function of synaptic connections. Hence these classes of genes are critical for brain plasticity. With DNA-chip technology it is possible to identify transcription responses to neural activity. These efforts have identified not only individual transcriptionally responsive genes but also gene networks responsible for brain plasticity.

[85] R. Scott, R. Bourtchuladze, S. Grossweiler et al. 2002. "CREB and the discovery of cognitive enhancers." *Journal of Molecular Neuroscience* 19, 171–177.

[86] T. Tully. 1997. "Regulation of gene expression and its role in long-term memory and synaptic plasticity." *Proceedings of the National Academy of Sciences* 94, 4239–4241. See also T. Tully, R. Bourtchouladze, R. Scott et al. 2003. "Targeting the CREB pathway for memory enhancers." *Nature Reviews: Drug Discovery* 2, 267–277.

[87] Glutamate is a neurotransmitter that induces the response of an AMPA receptor. This receptor in turn activates another protein (NMDA receptor), which admits calcium, bringing about the synaptic changes associated with learning.

role[88] in inducing synaptic changes that affect learning and memory encoding and consolidation, is in the pipeline. One such ampakine, Ampalex, has already been shown in preliminary tests to enhance cognitive performance.[89]

Medications already exist that have been shown to have enhancing effects. The antidepressants involved with the increase of serotonin and norepinephrine promote the production of neurotrophins, which are essential to keep the neurons healthy and are known to have growth-enhancing properties.[90] Nerve growth is believed to be enhanced by the effects of neurotrophin factor, which increases dendritic branching and spine density throughout the cortex.[91] Hence Prozac and similar medications have been shown not only to treat depression but also to create sometimes dramatic changes in awareness in those taking them.[92, 93] Since these medications have been used primarily for depression and anxiety, their cognition-enhancing qualities have only recently been considered. However, work has been progressing robustly on the role of norepinephrine reuptake inhibitors (NRIs), which are known to increase levels of both dopamine and norepinephrine in the frontal cortex, most particularly the dorsolateral prefrontal cortex.[94]

---

88 See D. Plotz. 2003. "Total Recall: The Future of Memory." Posted on slate.msn.com Tuesday, March 11, 2003.

89 "Open Your Mind." *Economist*, May 23, 2002.

90 See R. S. Duman, J. Malberg, S. Nakagawa et al. 2000. "Neuronal plasticity and survival in mood disorders." (Comment.) *Biological Psychiatry* 48, 732–739. See also L. Santarelli, M. Saxe, C. Gross et al. 2003. "Requirement of hippocampal neurogenesis for the behavioral effects of antidepressants." (Comment.) *Science* 301, 805–809.

91 B. Kolb and I. Q. Whishaw. 1996. *Fundamentals of Human Neuropsychology, 4th ed.* New York: Freeman. See also B. Kolb and I. Q. Whishaw. 1998. "Brain plasticity and behavior." *Ann Rev Psychol* 49, 43–64.

92 There seems to be some evidence that acetylcholinesterase inhibitors (such as Aricept and other cognitive enhancers) may similarly stimulate nerve growth.

93 P. Kramer. 1993. *Listening to Prozac: A Psychiatrist Explores Antidepressant Drugs and the Remaking of Self.* New York: Penguin.

94 S. M. Stahl. 2003. "Neurotransmission of Cognition, part 2. Selective NRIs are smart drugs: Exploiting regionally selective actions on both dopamine and norepinephrine to enhance cognition." *Journal of Clinical Psychiatry* 64, 110–111.

Dopamine, acetylcholine,[95] norepinephrine, and histamine have been shown to be important neurotransmitters in the frontal lobe.[96, 97] Other neurotransmitters are probably involved, but NRIs have taken the lead with the introduction of atomoxetine recently for the treatment of attention deficit/hyperactivity disorders (ADHD).[98] The cognitive-enhancing effects in the non-ADHD population have yet to be established, though the principle behind the creation of these drugs opens the opportunity for many more to be developed that will have enhancing effects.

Research is also being conducted on the enhancing effects of learning on cognition. Studies are showing that learning has a stimulating effect on neurotrophins, which can lead to neuron growth and improved cognition. In the near future the effects of various learning strategies on brain circuitry and cognition will likely be mapped out.[99]

---

[95] Medications for enhancing the neurotransmitter acetylcholine in the brain are being used to treat Alzheimer's disease. These medications, such as Aricept, are being studied regarding their possible benefit in enhancing memory and cognition in the normal population. See B. Evenson. "The guilty mind." *National Post*, February 8, 2003.

[96] Stimulation of the cholinergic system either pharmacologically or electrically has been shown to increase plasticity and improve learning. Alternatively, inactivation of this system may interfere with plasticity and forms of learning. See J. M. Edeline, 1999. "Learning-induced physiological plasticity in the thalamo-cortical sensory systems: A critical evaluation of receptive field plasticity, map changes and their potential mechanisms." *Prog. Neurobiol* 57, 165–224. Recent research has shown that stimulation of the basal forebrain cholinergic system affects cortical plasticity and learning. In keeping with this, lesions of the nucleus basalis impair skilled learning. The basal forebrain cholinergic system therefore plays an important role in guiding plasticity and learning. See M. Kilgard, 2003. "Cholinergic modulation of skill learning and plasticity." *Neuron* 38, 678–680.

[97] See S. M. Stahl. 2002. "Psychopharmacology of wakefulness: Pathways and neurotransmitters." *Journal of Clinical Psychiatry* 63, 551–552; see also, A. A. Schoups, R. C. Elliott, W. J. Friedman, and I. B. Black. 1995. "NGF and BDNF are differentially modulated by visual experience in the developing geniculocortical pathway." *Developmental Brain Research* 86, 326–334.

[98] Other medications with properties that block norepinephrine transporters in the frontal cortex include desipramine and reboxetine, as well as venlafaxine and many of the tricyclic antidepressants. For extensive discussion on this, see Stahl at footnote 94.

[99] A. A. Schoups, R. C. Elliott, W. J. Friedman, and I. B. Black. 1995. "NGF and BDNF are differentially modulated by visual experience in the developing geniculocortical pathway." *Developmental Brain Research* 86, 326–334.

Assessing Enhancement

Transcranial magnetic stimulation will have an important role in providing the framework for assessing enhancement from various factors, including learning, medications, electrical stimulation, stem cell stimulation, and others.[100] TMS has been shown to be capable of exciting as well as inhibiting areas of the brain. These capacities are allowing for the mapping of complex behaviors and the identification of sections of the brain that can potentially be stimulated to enhance cognitive functioning. In addition, TMS has been found to be useful in exploring the mechanisms and consequences of plasticity in the human cortex.[101] In examining regional reorganization, TMS can map patterns of connectivity within different cortical areas and their spinal projections, which is very useful in assessing, for example, the impact on the motor cortex of skill acquisition. TMS can also assist in evaluating the significance of behavior changes with the use of the "virtual lesion" capacity that allows for suppression of identified brain areas, thereby determining whether specific reorganization improves the function under study.

The long-term possibilities of TMS are encouraging. It has been shown to be useful in bringing about changes in cortical function. Short-term reorganization in the cortex has been achieved by means of repetitive TMS, whereby the magnitude and direction of the stimulation effectively induce plasticity depending on frequency, intensity, and total number of stimuli.[102] The functional changes induced involve interconnected cortical areas. The possibilities for long-term and even permanent changes remain to be explored. In the long term, TMS will

---

[100] New developments include "tensor diffusion imaging," which examines water molecules in brain nerve fiber. This technology, a modification of MRI, has the potential by measuring the speed and movement of diffusion to detect very small areas of function and abnormalities. See "Experimental brain scan improvement over MRI." 2003. *Schizophrenia Digest*, 1: 12–14.

[101] H. R. Siebner and J. Rothwell. 2003. "Transcranial magnetic stimulation: New insights into representational cortical plasticity." *Experimental Brain Research* 148, 1–16.

[102] Ibid.; footnote 101, p. 7.

likely be shown to produce changes in memory and complex behaviors.[103] This will include boosting normal cognitive as well as motor performance, as in skilled learning.[104]

PLASTICITY AND ENHANCEMENT

Recently two studies were reported in *Science* that hold major promise for our understanding of brain plasticity, as well as the possibilities for enhancement and treatment.[105] Two research teams, one at Cold Spring Harbor Laboratories and the other at New York University, reported on having taken detailed pictures of individual cerebral cortex neurons of anesthetized mice. They both used the technique of two-photon microscopy with laser pulses of infrared light that allowed them to study dendrites, the extended structures of neurons that receive and send stimuli. They were able to focus on the dendritic spines, which contain synapses, and showed that physical remodeling of the synapses occurs during learning. The technology for visualizing these changes was limited to one-half of a millimeter. However, methods are being worked on to extend this field deeper into the brain. Furthermore, given the progress of technology in the imaging field, it will soon be possible to visualize the human brain in its working state on the level of the micron.

The implications of this research are staggering with regards to understanding the process of plasticity in the smallest units of the brain.

---

103   Recent research on plasticity as it applies to the sensory cortices has shown that brain structure is unique to each individual. Essentially, this uniqueness relates to each individual's experiential history. Therefore, memory refers basically to the way events influence the brain and its future activity. See Siegel. 2001. "Memory: An overview with emphasis on developmental, interpersonal and neurobiological aspects." *Journal of the American Academy of Child and Adolescent Psychiatry* 40, 997–1011.

104   R. Topper, F. M. Mottaghy, M. Brugmann et al. 1998. "Facilitation of picture naming by focal transcranial magnetic stimulation of Wernicke's area." *Experimental Brain Research* 121, 371–378. See also W. Klimesch, P. Sauseng, and C. Gerloff. 2003. "Enhancing cognitive performance with repetitive transcranial magnetic stimulation at human individual alpha frequency." *European Journal of Neuroscience* 17, 1129–1133.

105   See G. Miller. 2003. "Spying on the brain, one neuron at a time." *Science* 300, 78–79.

Furthermore, the research offers the possibility of refinement in the mapping of the brain in terms of neural circuitry, and the functions of small, even one-neuron units. With this knowledge the door is open for very selective medications, which are in the developmental stage, and other technologies to inhibit or stimulate the brain and bring about further understanding of function and cognitive enhancement.[106] One alternative mechanism may be the use of microtransistors (or microchips) that would be guided to specific locations in the brain and attached to individual neurons or clusters of neurons to create permanent stimulation and, therefore, permanent enhancement of specific cognitive functions. For example, the placement of such a chip in the dorsal striatum or the nucleus accumbens, both of which are thought to be major pleasure centers, could inhibit or stimulate pleasure from certain actions, such as eating, or learning a set of facts, like a language. These centers are connected to the amygdala, which embeds specific events in memory.[107]

STEM CELLS AND ENHANCEMENT

Stem cell research will also be a rich area for advances in the long term. Knowledge will become available, perhaps through medications, TMS, or learning strategies, for inducing the conversion of stem cells in specific areas of the brain where enhancement of particular traits can be achieved. It should be possible in time to mobilize cells to specific sections of the brain, or even transplant stem cells from external sources.

THE FUTURE OF ENHANCEMENT

One day it might be possible to transplant sections of the brain of an embryo—or even more dramatic, of a dying person with enhanced memory or intelligence—into someone less fortunate. Research in these areas—stem cells and brain transplantation—therefore points to

---

[106] Enhancement is not limited to cognition per se. It will also involve enhancement of personality characteristics and motor skills.

[107] M. Blakeslee. "Madison Avenue and your brain." *Salon.* September 30, 2002.

the opportunity for correction of damaged areas of the brain, as well as enhancement. An alternative approach that is also likely in the future involves gene therapy. DNA that is known to be associated with specific learning, for example, could be introduced into brain cells to modify and enhance a person's ability to perform a specific function. Similarly, it may be possible to stimulate the genetic capacities of cells already known to perform specific functions to exact greater productivity.

Though futuristic, downloading of the brain may be also relevant in this context. With the possibility of downloading through modified techniques such as those described by Nicolelis at Duke, information from specific areas of the brain, or the whole brain, can be downloaded into a computer. Once downloaded, the information may be modified— for example, by adding a language capacity. Then the altered material may be uploaded into the same individual's brain. Nicolelis is working on achieving this return downloading of sensations from robot limbs that are being moved by the motor cortex of a subject. This return downloading may in time prove to be an excellent way to enhance cognitive skills.

Finally, it may become possible for the long term to manipulate the growth phase of the human brain, perhaps through the modification of genes. In their book *Liars, Lovers, and Heroes: What the New Brain Science Reveals About How We Become Who We Are*, Steven R. Quartz and Terrence J. Sejnowski comment on the fact that although chimpanzees have roughly 98.4 percent (some claim more) of the DNA of a human being, their frontal lobe—the seat of cognitive powers—is only one-sixth as large.[108] They speculate that the building of more complex structures may involve the same genes (referred to as homeobox genes—key developmental genes) in different ways, ways that affect how, when, and how long these genes are operative in the developmental process. They point out that the forebrain of the chimpanzee is nearly developed by two years of age, in part due to closure of the skull, whereas in the human less than one-half of the forebrain is developed by age two,

---

[108] S. R. Quartz and T. J. Sejnowski. 2002. *Liars, Lovers, and Heroes: What the New Brain Science Reveals About How We Become Who We Are.* New York: William Morrow, 41–59.

with the remainder developing after that, until the skull is fully formed. Hence, something about the timing of development—the influence of environment or culture—brings about a larger, more cognitively capable brain. In the long term, altering the timing of the completion of growth of the brain, perhaps through gene modification, thereby giving more time for culture to have an impact on development, may be a viable way to enhance cognition.

Neuroscience-Based Lie Detectors

The traditional lie detector, the polygraph, has been in existence for many years. It relies on physiological reactions—increased heart rate, respiration, and blood pressure, as well as sweating—as indicators that a person undergoing questioning about a set of events is fearful of getting caught, and is therefore lying. Though this test has been used in criminal investigations, critics of the polygraph insist that it can easily be defeated. Some people are very good at lying and are able to control their physiological responses, or they experience another sensation at the same time, which confuses the polygraph. Others who blur boundaries and readily incorporate happenings in their environment might see themselves as being responsible for acts they never committed, in which case their denial during a polygraph examination would result in a positive test.

In recent years, interest in detecting lying has grown because of the results of neuroscience research, which offers the possibility of producing highly accurate tests. The variety of tests devised to determine whether a person is lying includes near-infrared brain scans, thermal imaging, functional MRI, and Brain Fingerprinting®. The first of these, the near-infrared brain scan, is a test of blood flow. Devised by Dr. Britton Chance, a biophysicist at the University of Pennsylvania,[109] this method uses a headband containing near-infrared light emitters and detectors. Small changes in the prefrontal cortex, which is the site of decision making

---

[109] See R. James. "Brainwave monitoring becomes the ultimate lie detector." *SciScoop: Exploring Tomorrow,* January 6, 2003.

and is stimulated by deception, are read by sensors that, according to Chance, can detect changes that occur when a person makes a decision to lie, before the lie is actually articulated. Subjects are given a series of questions, some to be answered truthfully, others not. When a subject decides to lie, changes occur in the brain that are readily detected.[110] Though the device is still in the developmental stage, Chance claims that it will be capable in the near future of detecting covert activity in the prefrontal lobe.

Thermal imaging, using a heat-sensitive camera to detect increased blood flow around the eyes, is in the early developmental phase. Scientists working on this technique claim that when people lie, their eyes give off more heat than when they are telling the truth.[111] Others are also focusing on the face as a window to the brain, by using computers to analyze facial expressions, not heat, under the assumption that deception results in specific fleeting facial expressions that are usually imperceptible.[112] Detecting these is difficult even for a computer-regulated system. Furthermore, the relevant expressions must be shown to be statistically valid at a high degree of accuracy. Both thermal imaging and computer analysis of facial expressions may nonetheless be useful in the near future. Though unlikely to be 100 percent accurate—and nothing, including the testimony of witnesses, approaches this degree of accuracy—they may prove to be sufficiently accurate and of benefit at the very least in employee screening, if not in providing information about a person's character in court. Since near-infrared imaging seems further along in development and provides a direct assessment of the decision-making sections of the brain, it may be a more viable test for possible admissibility in court.

The use of functional MRI for lie detection is also currently being investigated. Dr. Daniel Langleben at the University of Pennsylvania has

---

[110] J. Loviglio. "New technology detects a lie before it's spoken: A peek at a brain can unmask a liar." *Pittsburgh Post-Gazette,* May 25, 2003.

[111] A. A. Moenssens. 2002. "Brain fingerprinting—can it be used to detect the innocence of persons charged with a crime?" *UMKC L. Rev. 70,* 891.

[112] P. Wen. 2001. "'Brain fingerprints' may offer better way to detect lying." *Boston Globe,* July 5, 2001.

been using fMRI to detect blood flow to specific sections of the brain. He conducted a study of 18 volunteers who were given specific playing cards and placed in an MRI scanner while undergoing interrogation by a computer that presented them with a specific card, and then asked them if they had it in their possession. When they lied, sections of their brain, particularly the anterior cingulate cortex and the superior frontal gyrus, lighted up more than when they told the truth. The anterior cingulate cortex, which has connections between the limbic system and the prefrontal cortex, is involved in emotional processing, decision making, and conflict resolution. This part of the brain is frequently activated when a lie is being told. On the other hand, it is involved with decision making generally, which is a confounding consideration. This section of the brain may be activated in a subject who is focusing on an anxiety-producing event that is unrelated to the issue in question. However, it appears that telling the truth does not create a distinctive brain print. Rather, there is diffuse activity in the temporal lobe, in contrast to lying, which seems to activate specific areas of the brain. More research must be conducted to ensure that the evaluations of fMRI patterns are highly specific to lying.[113]

BRAIN FINGERPRINTING

Perhaps the most developmentally advanced technology for lie detection is Brain Fingerprinting, which has already been admitted into evidence in one case,[114] though it has not gained wide acceptance in the scientific community. Brain Fingerprinting uses an EEG—a helmet of electrodes— to record event-related potentials induced by stimuli.[115] A subject is presented with words, phrases, or pictures while the EEG is recording his brain wave activity. Information acquired by the investigator that

---

[113] D. Langleben, L. Schroeder, J. Maldjian et al. 2002. "Brain activity during simulated deception: An event-related functional magnetic resonance study." *Neuroimage* 15, 727–732.

[114] *Harrington v. Iowa*, no. PCCV 073247 (Pottawattamie County D.C. Iowa, November 14, 2000).

ostensibly only the offender would know is presented to the subject. If he lacks knowledge of the information presented, the brain pattern will be uneventful. If, however, he does know the information and lies, then a specific brain wave, referred to as P300, is elicited. The P300 pattern is activated when the brain recognizes information (or a familiar object) as significant or surprising.[116] The object is to determine whether the person tested has the information that she denies stored in her brain.

Even though the inventor of Brain Fingerprinting claims an accuracy of nearly 100 percent, there are several problems with the technique.[117] First, the presence of drugs and alcohol can adversely affect the receiving and storing of information in the memory of the putative offender.[118] Second, the investigator has to have detailed information that only the subject would know, requiring much investigation. Currently, FBI and police reports are not so detailed. And finally, Brain Fingerprinting is not good for screening. Information must be known by the questioner for the test to work. The test could therefore not be used to randomly screen employees about their past.

In the short term, Brain Fingerprinting will be studied further and will most likely prove significantly useful in situations where unique factual information is available to investigators. Furthermore, with advances in behavioral genetics, it should be possible in the long term to correlate gene profiles with Brain Fingerprinting waves, thereby factoring out conditions such as generalized anxiety or fear, which may confound test results, and factoring in biological conditions, such as psychopathy, that are known to be highly associated with antisocial behavior. The relating of genetic information to Brain Fingerprinting patterns will enhance the statistical validity of the test results.

---

[115] "Investigative techniques—federal agency views on the potential application of 'Brain Fingerprinting.'" GAO Reports (RPT–Number: GAO–02–22), October 31, 2001.

[116] L. A. Farwell and E. Donchin. 1988. "Talking off the top of your head: Toward a mental prosthesis utilizing event-related brain potentials." Electroencephalographic and Clinical Neurophysiology 70, 510–523.

[117] B. Evenson. 2003. "The guilty mind." National Post, February 8, 2003.

[118] Ibid.; footnote 115, p. 8.

FUTURE RESEARCH

Refinements of fMRI will probably be made so that information can be detected with sensory input from the memory centers (amygdala and hippocampus) to the frontal cortex. This would probably include the observing of activation of the memory centers, the anterior cingulate cortex, and the superior frontal gyrus. A statistical study of tracking specific patterns of activation along the course of the impulse could be made. This might provide information on the relationship of the pattern to the degree and context of deception. Even more, through statistical examination of the pattern, it may be possible to determine the input of sections of the brain that are being activated during lying. An advanced MRI program that incorporates behavioral genetic data might be able to detect subtle changes in the hippocampus, for example, that would be translated into the content of feelings and meaning. Hence, it would be possible not only to determine whether the subject was lying but to derive some factual content on the nature of the lie.

Most likely, fMRI will be linked with transcranial magnetic stimulation to produce a powerful lie detection system.[119] TMS would be capable of blocking out or enhancing select portions of the brain. The blocking out of interference might improve the accuracy of fMRI in detecting activation in critical brain areas. Similarly, the enhancing of the memory centers, or anterior cingulate cortex, might result in greater sensitivity to lies that are camouflaged with confounding issues by the perpetrator. Also, fMRI will most likely be made smaller and portable so that it can be used relatively easily in a variety of settings, such as a police station, or by agencies such as the FBI and CIA.

Finally, as with the other areas of discussion, the future may bring the downloading of information from the brain. With the recognition that the mind is relegated to physical properties like biochemical and electrical

---

[119] SPECT and PET could have a role in detecting lies. Radiolabeled materials could be targeted to attach to receptors at the anterior cingulate gyrus or memory centers when these centers are activated by a lie. TMS could also act in this context as an inhibiting or activating measure. But the fact that radioactive material must be injected, and that PET requires a large room for its administration, makes fMRI a more likely candidate for development in lie detecting.

activity, downloading with the assistance of mathematical conversions is not beyond the realm of possibility. This would allow for information from decision-making centers—the frontal lobe and anterior cingulate cortex—and memory centers to be examined objectively to discover not only the presence of deception, but also important concealed information that may be highly relevant to a dispute or criminal case.

Potential Abuses of Neuroscience Information

Because neuroscience research is providing information about fundamental dimensions of our personality, most particularly intelligence and emotional capacity, it has the potential to powerfully influence every dimension of society. Major social institutions, from schools and courts to the workplace and the health care system, stand to benefit greatly from tests derived from this research. Tests have become a way of life in contemporary American culture. IQ and psychological tests have been the mainstay of schools, the workplace, and the military for well over 60 years. These tests serve many purposes for institutions.[120] To begin with, they are predictors of risks and success. For example, the Scholastic Aptitude Test for college admission is considered a good prognosticator of success or failure for academic work. Similarly, screening for drug use among prospective employees provides important information about the competency of such individuals to conduct skilled tasks. Personality tests that detail levels of emotional control, attitudes, and values frequently point out problem areas about potential employees in various working environments.

Tests provide a framework for assessing the potential costs associated with inclusion of people into a system. For example, genetic disorders have been estimated to be as high as 5 percent of all live births.[121] The costs associated with these disorders are high. It has been estimated that

---

[120]   See D. Nelkin and L. Tancredi. 1992. *Dangerous Diagnostics: The Social Power of Biological Information.* Chicago: University of Chicago Press, 9–14.

[121]   Ibid. at 66–68. See also D. J. Kevles. 1985. *In the Name of Eugenics.* New York: Alfred Knopf, 275–282.

they account for nearly 12 percent of adult admissions to hospitals and as much as 30 percent of pediatric admissions. Besides the classical genetic disorders, the full picture of a person's DNA will provide considerable information on that person's proneness to diseases. Knowledge of a person's genetic makeup would therefore be valuable for an employer who subsidizes health insurance, or an HMO in which patient profiling may result in exclusion of certain high-risk groups from coverage. Finally, tests point the way for remedial action on the part of social institutions. This has been most conspicuous in the school system, where tests establishing attention deficit disorders have justified requiring students to be treated with stimulants and other relevant medications. Similarly, IQ tests have been used to segregate intelligent children from those less fortunate and have resulted in specially designed educational programs for those intellectually handicapped. Recently, scientists at Stanford University have shown that "slow" children have a particular brain pattern. This opens up the question of whether screening should be undertaken to identify those with this pattern and to subject them to specialized educational programs to facilitate their remaining with the mainstream of students.[122]

Related to the objectives of tests is the strong tendency in the culture to reduce many problems to so-called solid objective criteria. This often takes the form of framing these problems in biological or medical terms.[123] Two benefits adhere to this type of reductionism. First, biological data are more concrete and can often be understood in terms of a number, as with enzyme assays; a chromosome description, as with genetic problems; and an image, as with biological abnormalities. Data such as these take on considerably greater weight with decision makers in social institutions. Yet frequently, insufficient attention is given to the problems of these tests, such as the inaccuracies of the techniques, possible misinterpretations, and the conceptual issue of whether they

---

[122] See E. Temple, G. K. Deutsch, R. A. Poldrack et al. 2003. "Neural deficits in children with dyslexia ameliorated by behavioral remediation: Evidence from functional MRI." *Proceedings of the National Academy of Sciences of the United States of America* 100, 2860–2865.

[123] For a discussion on this see footnote 120, pp. 37–50.

are actually measuring what they are claiming to measure. The second benefit is that biological tests suggest an easy course of action, which is to relegate those afflicted to the health care system. An institution can justifiably relieve itself of responsibility by pointing to the biological abnormality. We have seen in court cases that the display of a film of a person's brain abnormality has significantly greater weight than the testimony of a psychotherapist obtained from an interview.[124] The presence of a biological marker, whether an image or a blood test, is a powerful adjunct to a legal argument. On the other hand, by pointing to biological reasons, these tests relieve institutions, such as schools, of responsibility for critically examining their programs in search of problems that may explain or augment abnormalities.

As we move ahead in neuroscience, the possibility for exclusion of people from social benefits increases considerably. This is due in part to the fact that diagnostic methods are getting closer to what is actually occurring in the brain and are therefore viewed as more accurate than earlier tests. A fMRI or PET image is seen as a direct reflection of what is happening in the brain. This is in contradistinction to tests such as IQs that crudely measure performance, or traditional polygraphs that measure physiological responses. Tests that provide a direct examination of the functioning of the frontal lobe, hippocampus, or amygdala, for example, are probably going to be viewed with greater credibility. Hence, industry in the future will use scanning devices to screen employees much as they currently use genetic screening techniques to screen for particular chemical sensitivities.[125] Brain images will provide information about intelligence, skills, potential for learning, and emotional stability. At some stage in neuroscience development, it will be possible to screen the memory of a potential employee and thereby determine whether there is anything criminal or dissocial in his or her past.

Earlier I discussed comparisons of PET images with a test indicative of IQ. Combining genetic information with brain imaging will most likely provide considerable information about an individual's capacity

---

[124] See footnote 120, pp. 156–158.
[125] See footnote 120, pp. 95–101.

to learn and ultimately to work at certain jobs. Schools may require such tests early in a child's development and use them to profile intellectual abilities, personality characteristics, attentiveness, and motor skills. As a result of this early profiling, children may be channeled into educational directions based on the tests rather than their desires. College in some cases may consequently be precluded, which will inevitably limit career options. When and if downloading of the brain becomes feasible, a scan may allow for detailed examination of cognitive and personality characteristics. Not only will this allow for discrimination based on a test, but it offers the possibility of egregious abuse of an individual's privacy. Thoughts, feelings, desires, values, and the full range of emotional characteristics would conceivably be available for review and would be the basis of exclusion from an academic program or consideration for a job. People generally have a wide range of feelings and thoughts—for example, a person may want to kill someone during a moment of intense anger—that are never actualized but may serve as reasons for a decision against hiring. Where improvement to the norm (or even enhancement) is possible through computer programs, the downloading may empower social institutions to insist that the individual accept corrections and additions to his intelligence and personality, which can be achieved through uploading of altered material. This would be tampering with the individual's very identity.

The possibility of enhancement also offers the opportunity for other abuse and discrimination. Medications and other techniques, which will probably be in short supply during initial development, will most likely be limited to those who have sufficient financial resources or who are deemed worthy because they have been found to be intelligent enough through prior testing to benefit the most from enhancement. Most likely a range of factors will be taken into consideration, such as the age of a prospective candidate, his or her history, and basic personality characteristics. Even if these benefits were available to nearly everyone, the outcome would probably entail more than just a ratcheting up of what constitutes a normal IQ, or acceptable moral behavior. Inevitably, societal trade-offs accompany any treatment or enhancement effort. Questions like the following emerge: What if anything will be diminished secondary to enhancement? What impact would enhancement have on the whole personality and the range of types of personalities in society? Would we be willing to have a larger percentage of less productive members or perhaps of antisocial personalities, knowing that the "normal" has been elevated to a higher level? Would it be ethical to require people to have their brain chemically and mechanically altered? At what point do the requirements of society for a superior workforce have precedence over the desires of the individual? Finally, successful enhancement might be used to create a superior group in society that could politically and financially dominate the remaining groups. This would create a biological underclass of people—not highly educated, possibly uninsured for health benefits, and relatively unemployable except in unskilled jobs.

Neuroscience-based tests for determining the veracity of a person's statements also present powerful opportunities for abuse. The traditional lie detector, the polygraph, is generally considered not to be highly trustworthy as a test for lying. In fact, it has been considered unreliable for admissibility into evidence regarding the truthfulness of a statement by a defendant or witness. Nonetheless, some industries and federal agencies use the test to assess job applicants and screen workers.[126] Brain Fingerprinting, near-infrared imaging, and fMRI, as they develop over the next several years, will probably become increasingly reliable measures of lying, possibly to the point of being admissible as evidence in a criminal case or civil dispute. The developer of Brain Fingerprinting claims that his test is already highly accurate. Industry, the federal government, schools, and law enforcement agencies are likely to rely increasingly on these tests, which will be used to screen employees; determine the acceptability of students into specialized educational programs, including higher education; and be the basis for suspicion regarding a criminal act. Should "brain" technologies reach the stage in the distant future when they are able to download the memory centers or the frontal

---

[126] See footnote 107.

lobe and anterior cingulate cortex of the brain, social institutions will not only have information about an individual's behavior in the past, but also be privy to many aspects of his or her privacy that are irrelevant to their legal concerns. Such information may be very useful in providing leads for solving a criminal case, but it is unlikely ever to become 100 percent accurate because of potential error in technique and interpretation. Unfortunately, there is a chance that the small margin of inaccuracy and unreliability will be ignored in the interests of administrative simplicity, which means that some individuals will be falsely accused of crimes or inappropriately denied such social benefits as employment and higher education.

At the rate neuroscience research is progressing, there is no doubt that major advances will be made over the next decade. Many, though perhaps not all, of the potential developments discussed here are likely to come about. Though great benefits will be enjoyed by society—through treating Alzheimer's disease and other cognitive impairments, discovering measures for enhancing human potential, and developing reliable lie detectors involving the brain—there is the ever present danger that institutions will misuse this powerful information. It will be important that all members of society profit equally from these neuroscience achievements and that the opportunity for abuse and delegitimation of less fortunate groups be avoided at all costs.

8

Prediction, Litigation, Privacy, and Property

Some Possible Legal and Social Implications of Advances in Neuroscience

HENRY T. GREELY[1]
Deane F. and Kate Edelman Johnson Professor of Law;
professor, by courtesy, of genetics, Stanford University

"There's no art/To find the mind's construction in the face:/He was a gentleman on whom I built/An absolute trust."[2]

THE LAMENT OF DUNCAN, King of Scotland, for the treason of the Thane of Cawdor, his trusted nobleman, echoes through time as we continue to feel the sting of not *knowing* the minds of those people with whom we deal. From "We have a deal" to "Will you still love me tomorrow?" we continue to live in fundamental uncertainty about the minds of others. Duncan

---

[1] I want to thank particularly my colleagues John Barton, George Fisher, and Tino Cuellar for their helpful advice on intellectual property, evidentiary issues, and neuroscience predictions in the criminal justice system, respectively. I also want to thank the participants at the AAAS/Dana Foundation workshop on law and neuroscience for their useful comments, as well as colleagues and students who made suggestions at talks on these topics I gave at Stanford during the fall of 2003. Last, but not least, I want to thank my research assistant, Melanie Blunschi, for her able help.

[2] Shakespeare, *Macbeth*, act 1, scene 4, lines 11–14.

himself emphasized this by immediately giving his trust to Cawdor's conqueror, one Macbeth, with fatal consequences. But at least some of this uncertainty may be about to lift, for better or for worse.

Neuroscience is rapidly increasing our knowledge of the functioning, and malfunctioning, of that intricate three-pound organ, the human brain. In expanding our understanding of something so central to human existence, science is necessarily provoking changes in both our society and its laws. This paper seeks to forecast and explore the social and legal changes that neuroscience might bring in four areas: prediction, litigation, confidentiality and privacy, and patents. It complements the paper in this volume written by Stephen Morse, which covers issues of personhood and responsibility, informed consent, the reform of existing legal doctrines, enhancement of normal brain functions, and the admissibility of neuroscience evidence.

Two notes of caution are in order. First, this paper may appear to paint a gloomy picture of future threats and abuses. In fact, the technologies discussed seem likely to have benefits far outweighing their harms. It is the job of people looking for ethical, legal, and social consequences of new technologies to look disproportionately for troublesome consequences—or, at least, that's the convention. Second, as Niels Bohr (probably) said, "It is always hard to predict things, especially the future."[3] This paper builds on experience gained in studying the ethical, legal, and social implications of human genetics over the last decade. That experience, for me and for the whole field, has included both successes and failures. In neuroscience, as in genetics, accurately envisioning the future is particularly difficult, as one must foresee successfully both what changes will occur in the science and how they will affect society. I am confident about only two things concerning this paper: first, it discusses at length some things that will never happen; and second, it ignores what will prove to be some of the most important social and legal implications of neuroscience. Nonetheless, I hope the paper will be useful as a guide to beginning to think about these issues.

---

[3] The source of this common saying is surprisingly hard to pin down, but Bohr seems the most plausible candidate. See Henry T. Greely, *Trusted Systems and Medical Records: Lowering Expectations*, Stan. L. Rev. 52: 1585, 1591 n. 9 (2000).

Prediction

Advances in neuroscience may well improve our ability to make predictions about an individual's future. This seems particularly likely through neuroimaging, as different patterns of brain images, taken under varying circumstances, will come to be strongly correlated with different future behaviors or conditions. The images may reveal the *structure* of the living brain, through technologies such as computed axial tomography (CAT) scans or magnetic resonance imaging (MRI), or they may show how different parts of the brain *function*, through positron-emission tomography (PET) scans, single-photon-emission computed tomography (SPECT) scans, or functional magnetic resonance imaging (fMRI).

Neuroscience might make many different kinds of predictions about people. It might predict, or reveal, mental illness, behavioral traits, and cognitive abilities, among other things. For the purposes of this paper, I have organized these predictive areas not by the nature of the prediction but by who might use the predictions: the health care system, the criminal justice system, schools, businesses, and parents.

That new neuroscience methods are used to make predictions is not necessarily good or bad. Our society makes predictions about people all the time: when a doctor determines a patient's prognosis, when a judge (or a legislature) sentences a criminal, when colleges use the Scholastic Aptitude Test, and when automobile liability insurers set rates. But although prediction is common, it is not always uncontroversial.

THE ANALOGY TO GENETIC PREDICTIONS

The issues raised by predictions based on neuroscience are often similar to those raised by genetic predictions. Indeed, in some cases the two areas are the same—genetic analysis can powerfully predict several diseases of the brain, including Huntington's disease and some cases of early-onset Alzheimer's disease. Experience of genetic predictions teaches at least three important lessons.

First, a claimed ability to predict may not, in fact, exist. Many associations between genetic variations and various diseases have been claimed, only to fail the test of replication. Interestingly, many of these failures have involved two mental illnesses, schizophrenia and bipolar disorder.

Second, and more important, the strength of the predictions can vary enormously. For some genetic diseases, prediction is overwhelmingly powerful. As far as we know, the only way a person with the genetic variation that causes Huntington's disease can avoid dying of that disease is to die first from something else. On the other hand, the widely heralded "breast cancer genes," BRCA 1 and BRCA 2, though they substantially increase the likelihood that a woman will be diagnosed with breast or ovarian cancer, are not close to determinative. Somewhere between 50 and 85 percent of women born with a pathogenic mutation in either of those genes will get breast cancer; 20 to 30 percent (well under half) will get ovarian cancer. Men with a mutation in BRCA 2 have a hundred-fold greater risk of breast cancer than average men—but their chances are still under 5 percent. A prediction based on an association between a genetic variation and a disease, even when true, can be very strong, very weak, or somewhere in between. The popular perception of genes as extremely powerful is probably a result of ascertainment bias: the diseases first found to be caused by genetic variations were examples of very powerful associations *because* powerful associations were the easiest to find. If, as seems likely, the same holds true for predictions from neuroscience, such predictions will need to be used very carefully.

Finally, the use of genetic predictions has proven controversial, both in medical practice and in social settings. Much of the debate about the uses of human genetics has concerned its use in predicting the health or traits of patients, insureds, employees, fetuses, and embryos. Neuroscience seems likely to raise many similar issues.

HEALTH CARE

Much of health care is about prediction—predicting the outcome of a disease, predicting the results of a treatment for a disease, predicting the risk of getting a disease. When medicine, through neuroscience, genetics, or other methods, makes an accurate prediction that leads to a useful intervention, the prediction is clearly valuable. But predictions also can cause problems when they are inaccurate (or are perceived inaccurately by patients). Even if the predictions are accurate, they still have uncertain value if no useful interventions are possible. These problems may justify regulation of predictive neuroscientific medical testing.

Some predictive tests are inaccurate, either because the scientific understanding behind them is wrong or because the test is poorly performed. In other cases the test may be accurate in the sense that it gives an accurate assessment of the probability of a certain result, but any individual patient may not have the most likely outcome. In addition, patients or others may misinterpret the test results. In genetic testing, for example, a woman who tests positive for a BRCA 1 mutation may believe that a fatal breast cancer is inevitable, when, in fact, her lifetime risk of breast cancer is between 50 and 85 percent, and her chance of dying from a breast cancer is roughly one-third of the risk of diagnosis. Alternatively, a woman who tests negative for the mutation may falsely believe that she has no risk for breast cancer and might stop breast self-examinations or mammograms to her harm. Even very accurate tests may not be very useful. Genetic testing to predict Huntington's disease is quite accurate, yet, with no useful medical interventions, a person may find foreknowledge of Huntington's disease not only unhelpful but psychologically or socially harmful. These concerns have led to widespread calls for regulation of genetic testing.[4]

The same issues can easily arise in neuroscience. Neuroimaging, for example, might easily lead to predictions, with greater or lesser accuracy, that

4   See, for example, Secretary's Advisory Committee on Genetic Testing, *Enhancing the Oversight of Genetic Tests: Recommendations of the SACGT*, National Institutes of Health (July 2000), report available at http://www4.od.nih.gov/oba/sacgt/reports/oversight_report.htm; N. A. Holtzman and M. S. Watson, eds., *Promoting Safe and Effective Genetic Testing in the United States: Final Report of the Task Force on Genetic Testing* (Baltimore: Johns Hopkins University Press, 1997); and Barbara A. Koenig, Henry T. Greely, Laura McConnell, Heather Silverberg, and Thomas A. Raffin, "PGES Recommendations on Genetic Testing for Breast Cancer Susceptibility." *Journal of Women's Health* (June 1998) 7: 531–545.

an individual will be diagnosed with one of a variety of neurodegenerative diseases. Such imaging tests may be inaccurate, may present information patients find difficult to evaluate, and may provide information of dubious value and some harm. One might want to regulate some such tests along the lines proposed for genetic tests: proof that the test was effective at predicting the condition in question, assessment of the competency of those performing the test, required informed consent so that patients appreciate the test's possible consequences, and mandatory post-test counseling to ensure that patients understand the results.

The Food and Drug Administration (FDA) has statutory jurisdiction over the use of drugs, biologicals, and medical devices. It requires proof that covered products are both safe and effective. The FDA has asserted that it has jurisdiction over genetic tests as medical devices, but it has chosen to impose significant regulation only on genetic tests sold by manufacturers as kits to clinical laboratories, physicians, or consumers. Tests done as "home brews" by clinical laboratories have been subject only to very limited regulation, which does not include proof of safety or efficacy. Neuroscience tests might well be subject to even less FDA regulation. If the test used an existing, approved medical device, such as an MRI machine, no FDA approval of this additional use would be necessary. The test would be part of the "practice of medicine," expressly not regulated by the FDA.

The FDA also implements the Clinical Laboratory Improvement Amendments Act (CLIA), along with the Centers for Disease Control and Prevention and the Centers for Medicare and Medicaid Services. CLIA sets standards for the training and working conditions of clinical laboratory personnel and requires periodic testing of laboratories' proficiency at different tests. Unless the tests were done in a clinical laboratory—through, for example, pathological examination of brain tissue samples or analysis of chemicals from the brain—neuroscience testing would also seem to avoid regulation under CLIA.

At present, neuroscience-based testing, particularly through neuroimaging using existing (approved) devices, seems to be entirely unregulated except, to a very limited extent, by malpractice law. One important policy question should be whether to regulate such tests, through government action or by professional self-regulation.

CRIMINAL JUSTICE

The criminal justice system makes predictions about individuals' future behavior in sentencing, parole, and other decisions, such as civil commitment for sex offenders.[5] The trend in recent years has been to limit the discretion of judges and parole boards in using predictions by setting stronger sentencing guidelines and mandatory sentences. Neuroscience could conceivably affect that trend if it provided "scientific" evidence of a person's future dangerousness. Such evidence might be used to increase sentencing discretion—or it might provide yet another way to limit such discretion.[6]

One can imagine neuroscience tests that could predict a convicted defendant was particularly likely to commit dangerous future crimes by showing that he has, for example, poor control over his anger, his aggressiveness, or his sexual urges. This kind of evidence has been used in the past; neuroscience may come up with ways that either are more accurate or that *appear* more accurate (or more impressive). For example, two papers[7] have already linked criminality to variations in the gene for monoamine oxidase A, a protein that plays an important role in the

---

[5]   Prosecutors also make predictions in using their discretion in charging crimes and in plea bargaining; the police also use predictions in deciding on which suspects to focus. My colleague Tino Cuellar pointed out to me that neuroscience data, from a current prosecution or investigation of an individual or from earlier investigations of that person, might play a role in deciding the criminal charge or whether to plea bargain.

[6]   The implications of neuroscientific assessments of a person's state of mind at the time of the crime for criminal liability are discussed in Professor Morse's paper. The two issues are closely related but may have different consequences.

[7]   See H. G. Brunner, M. Nelen, X. O. Breakefield, H. H. Ropers, and B. A. van Oost, "Abnormal Behavior Associated with a Point Mutation in the Structural Gene for Monoamine Oxidase A." *Science* 262: 5133–5136 (October 22, 1993), discussed in V. Morrell "Evidence Found for a Possible 'Aggression' Gene." *Science* 260: 1722–1724 (June 18, 1993); and Avshalon Caspi, Joseph McClay, Terrie E. Moffitt, Jonathan Mill, Judy Martin, Ian W. Craig, Alan Taylor, and Richie Poulton, "Role of Genotype in the Cycle of Violence in Maltreated Children." *Science* 297: 851–854 (August 2, 2002), discussed in Erik Stokstad, "Violent Effects of Abuse Tied to Gene." *Science* 297: 752 (August 2, 2002).

brain. Genetic tests may seem more scientific and more impressive to a judge, jury, or parole board than a psychologist's report. The use of neuroscience to make these predictions raises at least two issues: Are neuroscience tests for future dangerousness or lack of self-control valid at all? If so, how accurate do they need to be before they should be used?

The law has had experience with claims that inherent violent tendencies can be tested for. The XYY syndrome was widely discussed and accepted, in the literature though not by the courts,[8] in the late 1960s and early 1970s. Men born with an additional copy of the Y chromosome were said to be much more likely to become violent criminals. Further research revealed, about a decade later, that XYY men were somewhat more likely to have low intelligence and to have long arrest records, typically for petty or property offenses. They did not have a higher than average predisposition to violence.

If, unlike XYY syndrome, a tested condition were shown to predict reliably future dangerousness or lack of control, the question would then become how accurate the test must be in order for it to be used. A test of dangerousness or lack of control that was only slightly better than flipping coins should not be given much weight; a perfect test could be. At what accuracy level should the line be set?

The Supreme Court has recently spoken twice on the civil commitment of sexual offenders, both times reviewing a Kansas statute.[9] The Kansas act authorizes civil commitment of a "sexually violent predator," defined as "any person who has been convicted of or charged with a sexually violent offense and who suffers from a mental abnormality or personality disorder which makes the person likely to engage in repeat acts of sexual violence."[10] In *Kansas v. Hendricks*, the

---

[8] See the discussion of the four unsuccessful efforts to use XYY status as a defense in criminal cases in Deborah W. Denno, "Human Biology and Criminal Responsibility: Free Will or Free Ride?" 137 U. PA. L. REV. 613, 620–22 (1988).

[9] See two excellent recent discussions of these cases: Stephen J. Morse, *Uncontrollable Urges and Irrational People*, VA. L. REV. 88 (2002): 1025; and Peter C. Pfaffenroth, *The Need for Coherence: States' Civil Commitment of Sex Offenders in the Wake of Kansas v. Crane*, STAN. L. REV. 55 (2003): 2229.

[10] Kan. Stat. Ann. §59–29a02(a) (2003).

Court held the state law constitutional against a substantive due process claim because it required, in addition to proof of dangerousness, proof of the defendant's lack of control. "This admitted lack of volitional control, coupled with a prediction of future dangerousness, adequately distinguishes Hendricks from other dangerous persons who are perhaps more properly dealt with exclusively through criminal proceedings."[11] It held that Hendricks's commitment survived attack on ex post facto and double jeopardy grounds because the commitment procedure was neither criminal nor punitive.[12]

Five years later, the Court revisited this statute in *Kansas v. Crane*.[13] It held that the Kansas statute could be applied constitutionally only if there were a determination of the defendant's lack of control and not just proof of the existence of a relevant "mental abnormality or personality disorder":

> It is enough to say that there must be proof of serious difficulty in controlling behavior. And this, when viewed in light of such features of the case as the nature of the psychiatric diagnosis, and the severity of the mental abnormality itself, must be sufficient to distinguish the dangerous sexual offender whose serious mental illness, abnormality, or disorder subjects him to civil commitment from the dangerous but typical recidivist convicted in an ordinary criminal case.[14]

We know then that, at least in civil commitment cases related to prior sexually violent criminal offenses, proof that the particular defendant had limited power to control his actions is constitutionally necessary. There is no requirement that this evidence, or proof adduced in sentencing or parole hearings, convince the trier of fact beyond a reasonable doubt. The Court gives no indication of how strong that evidence must be or how its scientific basis would be established. Would any evidence that passed *Daubert* or *Frye* hearings be sufficient for civil commitment (or

---

[11]   521 U.S. 346, 360 (1997).

[12]   Ibid.

[13]   534 U.S. 407 (2002).

[14]   Ibid., p. 413.

for enhancing sentencing or denying parole), or would some higher standard be required?

It is also interesting to speculate on how evidence of the accuracy of such tests would be collected. It is unlikely that a state or federal criminal justice system would allow a randomized double-blind trial, performing the neuroscientific dangerousness or volition tests on all convicted defendants at the time of their conviction and then releasing them to see which ones would commit future crimes. That judges, parole boards, or legislatures would insist on rigorous scientific proof of connections between neuroscience evidence and future mental states seems doubtful.

SCHOOLS

Schools commonly use predictions of individual cognitive abilities. Undergraduate and graduate admissions are powerfully influenced by applicants' scores on an alphabet's worth of tests: ACT, SAT, LSAT, MCAT, and GRE, among others. Even those tests, such as the MCAT, that claim to test knowledge rather than aptitude, use the applicant's tested knowledge as a predictor of her ability to function well in school, either because she has that background knowledge or because her acquisition of the knowledge demonstrates her abilities. American primary and secondary education uses aptitude tests less frequently, although some tracking does go on. And almost all of those schools use grading (after a certain level), which the school or others—such as other schools, employers, and parents—can use to make predictions.

It is conceivable that neuroscience could provide other methods of testing ability or aptitude. Of course, the standard questions of the accuracy of those tests would apply. Tests that are highly inaccurate usually should not be used. But even assuming the tests are accurate, they would raise concerns. While they might be used only positively—as Dr. Alfred Binet intended his early intelligence test to be used, to identify children who need special help—to the extent that they were used to deny students, especially young children, opportunities, they would seem more troubling.

It is not clear why a society that uses aptitude tests so commonly for admission into elite schools should worry about the tests' neuroscience equivalents. The SAT and similar aptitude tests claim that student preparation or effort will not substantially affect the tests' results, just as, presumably, preparation (at least in the short term) seems unlikely to alter neuroscience tests of aptitude. The existing aptitude tests, though widely used, remain controversial. Neuroscience tests, particularly if given and acted upon at an early age, are likely to exacerbate the discomfort we already feel with predictive uses of aptitude tests in education.

BUSINESSES

Perhaps the most discussed social issue in human genetics has been the possible use—or abuse—of genetic data by businesses, particularly insurers and employers. Most, but not all, commentators have favored restrictions on the use of genetic information by health insurers and employers.[15] And legislators have largely agreed. Over 45 states and, to some extent, the federal government restrict the use of genetic information in health insurance. Eleven states impose limits on the use of genetic information by life insurers, but those constraints are typically weak. About 30 states limit employer-ordered genetic testing or the use of genetic information in employment decisions, as does, to some very unclear extent, the federal government, through the Americans with Disabilities Act.[16] And 2004 might be the year when broad federal legislation against "genetic

---

[15] For a representative sample of views, see Kathy L. Hudson, Karen H. Rothenberg, Lori B. Andrews, Mary Jo Ellis Kahn, and Francis S. Collins, "Genetic Discrimination and Health Insurance: An Urgent Need for Reform." Science 270 (1995): 391 (broadly favoring a ban on discrimination); Richard A. Epstein, The Legal Regulation of Genetic Discrimination: Old Responses to New Technology, B.U.L. REV. 74 (1994): 1 (opposing a ban on the use of genetic information in employment discrimination); Henry T. Greely, Genotype Discrimination: The Complex Case for Some Legislative Protection, U. PA. L. REV. 149 (2001): 1483 (favoring a carefully drawn ban, largely to combat exaggerated fears of discrimination); and Colin S. Diver and Jane M. Cohen, Genophobia: What Is Wrong with Genetic Discrimination? U. PA. L. REV. 149 (2001): 1439 (opposing a ban on its use in health insurance).

[16] For the most up-to-date information on state law in this area, see Ellen W. Clayton, "Ethical, Legal, and Social Implications of Genomic Medicine." New Eng. J. Med. 349 (2003): 542.

discrimination" is finally passed.[17] Should similar legislation be passed to protect people against "neuroscience" discrimination?

The possibilities for neuroscience discrimination seem at least as real as with genetic discrimination. A predictive test showing that a person has a high likelihood of developing schizophrenia, bipolar disorder, early-onset Alzheimer's disease, early-onset Parkinson's disease, or Huntington's disease could certainly provide insurers or employers with an incentive to avoid that person. To the extent one believes that health coverage should be universal or that employment should be denied or terminated only for good cause, banning "neuroscientific discrimination" might be justified as an incremental step toward this good end. Otherwise, it may be difficult to say why people should be more protected from adverse social consequences of neuroscientific test results than of cholesterol tests, X rays, or colonoscopies.

Special protection for genetic tests has been urged on the ground that genes are more fundamental, more deterministic, and less the result of personal actions or chance than other influences on health. Others have argued against such "genetic exceptionalism," denying special power to genes and contending that special legislation about genetics only confirms in the public a false view of genetic determinism. Still others, including me, have argued that the public's particularly strong fear of genetic test results, even though exaggerated, justifies regulation in order to gain concrete benefits from reducing that fear. The same arguments could be played out with respect to predictive neuroscience tests. Although this is an open empirical question, it does seem likely that the public's perception of the fundamental or deterministic nature of genes does not exist with respect to neuroscience.

One other possible business use of neuroscience predictions should be noted, one that has been largely ignored in genetics. Neuroscience

---

[17]   After considering, but not adopting, similar legislation since 1997, the Senate in October 2003 passed the Genetic Information NonDiscrimination Act, S. 1053. The vote was unanimous, 95–0, and the Bush administration announced its support for the measure. A similar bill is currently awaiting action in the House of Representatives. See Aaron Zitner, "Senate Blocks Genetic Discrimination," *Los Angeles Times*, October 15, 2003, sec. 1, p. 16.

might be used in marketing. Firms might use neuroscience techniques on test subjects to enhance the appeal of their products or the effectiveness of their advertising. Individuals or focus groups could, in the future, be examined under fMRI. At least one firm, Brighthouse Institute for Thought Sciences, has embraced this technology and, in a press release from 2002, announced "its intentions of revolutionizing the marketing industry."[18]

More alarmingly, if neuromonitoring devices were perfected that could study a person's mental function without his or her knowledge, information intended to predict a consumer's preferences might be collected for marketing purposes. Privacy regulation seems appropriate for the undisclosed monitoring in the latter example. Regulating the former seems less likely, although it might prove attractive if such neuroscience-enhanced market research proved too effective an aid to selling.

PARENTS

The prenatal use of genetic tests to predict the future characteristics of fetuses, embryos, or as-yet-unconceived offspring is one of the most controversial and interesting issues in human genetics. Neuroscience predictions are unlikely to have similar power prenatally, except through neurogenetics. It is possible that neuroimaging or other nongenetic neuroscience tests might be performed on a fetus during pregnancy. Structural MRI has been used as early as about 24 weeks to look for major brain malformations, following up on earlier suspicious sonograms. At this point, no one appears to have done fMRI on the brain of a fetus; the classic method of stimulating the subject and watching which brain regions react would be challenging in utero, though not necessarily impossible. In any event, fetal neuroimaging seems likely to give meaningful results only for serious brain problems and even then at a fairly late stage of fetal development so that the

---

[18] "Brighthouse Institute for Thought Sciences Launches First 'Neuromarketing' Research Company," press release (June 22, 2002) found at http://www.prweb.com/releases/2002/6/prweb40936.php.

most plausible intervention, abortion, would be rarely used and only in the most extreme cases.[19]

Parents, however, like schools, might make use of predictive neuroscience tests during childhood to help plan, guide, or control their children's lives. Of course, parents already try to guide their children's lives, based on everything from good data to wishful thinking about a child's abilities. Would neuroscience change anything? It might be argued that parents would take neuroscience testing more seriously than other evidence of a child's abilities because of its scientific nature, and thus perhaps exaggerate its accuracy. More fundamentally, it could be argued that, even if the test predictions were powerfully accurate, too extreme parental control over a child's life is a bad thing. From this perspective, any procedures that are likely to add strength to parents' desire or ability to exercise that control should be discouraged. On the other hand, society vests parents with enormous control over their children's upbringing, intervening only in strong cases of abuse. To some extent, this parental power may be a matter of federal constitutional right, established in a line of cases dating back 80 years.[20]

This issue is perhaps too difficult to be tackled. It is worth noting, though, that government regulation is not the only way to approach it. Professional self-regulation, insurance coverage policies, and parental education might all be methods to discourage any perceived overuse of children's neuroscience tests by their parents.

Litigation Uses

Predictions may themselves be relevant in some litigation, particularly the criminal cases discussed above, but other, nonpredictive uses of neuroscience might also become central to litigated cases. Neuroscience *might* be able to provide relevant, and possibly determinative, evidence of a witness's mental

---

[19] It seems conceivable that MRI results of a fetal brain might ultimately be used in conjunction with prenatal neurosurgery.

[20] See, for example, *Pierce v. Society of Sisters*, 268 U.S. 510 (1925); *Meyer v. Nebraska*, 262 U.S. 390 (1923).

state at the time of testimony, ways of eliciting or evaluating a witness's memories, or other evidence relevant to a litigant's claims. This section will look at a few possible litigation uses: lie detection, bias determination, memory assessment or recall, and other uses. Whether any of these uses is scientifically possible remains to be seen. It is also worth noting that the extent of the use of any of these methods will also depend on their cost and intrusiveness. A method of, for example, truth determination that required an intravenous infusion or examination inside a full-scale MRI machine would be used much less than a simple and portable headset.

The implications of any of these technologies for litigation seem to depend largely on four evidentiary issues. First, will the technologies pass the *Daubert*[21] or *Frye*[22] tests for the admissibility of scientific evidence? (I leave questions of *Daubert* and *Frye* entirely to Professor Morse.) Second, if they are held sufficiently scientifically reliable to pass *Daubert* or *Frye*, are there other reasons to forbid or to compel the admissibility of the results of such technologies when used voluntarily by a witness? Third, would the refusal—or the agreement—of a witness to use one of these technologies itself be admissible in evidence? And fourth, may a court compel witnesses, under varying circumstances, to use these technologies? The answers to these questions will vary with the setting (especially criminal or civil), with the technology, and with other circumstances of the case, but they provide a useful framework for analysis.

DETECTING LIES OR COMPELLING TRUTH

The concept behind current polygraph machines dates back to the early twentieth century.[23] They seek to measure various physiological reactions

---

[21] *Daubert v. Merrell Dow Pharmaceuticals*, 516 U.S. 869; 116 S. Ct. 189; 133 L. Ed. 2d 126 (1993).

[22] *Frye v. United States*, 54 App. D.C. 46, 293 F. 1013 (1923, D.C. Cir.).

[23] A National Academy of Sciences panel examining polygraph evidence dated the birth of the polygraph machine to William Marston, between 1915 and 1921. Committee to Review the Scientific Evidence on the Polygraph, National Research Council, *The Polygraph and Lie Detection* at 291–97 (Mark H. Moore and Anthony A. Braga, eds., 2003). Marston was the polygraph examiner whose testimony was excluded in *Frye v. United States*.

associated with anxiety, such as sweating, breathing rate, and blood pressure, in the expectation that those signs of nervousness correlate with the speaker's knowledge that what he is saying is false. American courts have generally, but not universally, rejected them, although they are commonly used by the federal government for various security clearances and investigations.[24] It has been estimated that their accuracy is about 85 to 90 percent.[25]

Now imagine that neuroscience leads to new ways to determine whether or not a witness is telling a lie or even to compel a witness to tell the truth. A brain-imaging device might, for example, be able to detect patterns or locations of brain activity known from experiments to be highly correlated with the subject's consciousness of falsehood. (I will refer to this as "lie detection.") Alternatively, drugs or other

---

[24] See the discussion in *United States v. Scheffer*, 523 U.S. 303, 310–11 (1998). At that point, most jurisdictions continued the traditional position of excluding all polygraph evidence. Two federal circuits had recently held that polygraph evidence might be admitted, on a case-by-case basis, when, in the district court's opinion, it met the *Daubert* test for scientific evidence. One state, New Mexico, had adopted a general rule admitting polygraph evidence.

[25] Justice Stevens characterized the state of the scientific evidence as follows in his dissent in *United States v. Scheffer*:

> There are a host of studies that place the reliability of polygraph tests at 85 percent to 90 percent. While critics of the polygraph argue that accuracy is much lower, even the studies cited by the critics place polygraph accuracy at 70 percent. Moreover, to the extent that the polygraph errs, studies have repeatedly shown that the polygraph is more likely to find innocent people guilty than vice versa. Thus, exculpatory polygraphs—like the one in this case—are likely to be more reliable than inculpatory ones.

United States v. Scheffer, 523 U.S. 303, 333 (1998) (Stevens, J., dissenting) (footnotes omitted).

A committee of the National Academy of Sciences has recently characterized the evidence as follows:

> Notwithstanding the limitations of the quality of the empirical research and the limited ability to generalize to real-world settings, we conclude that in populations of examinees such as those represented in the polygraph research literature, untrained in countermeasures, specific-incident polygraph tests can discriminate lying from truth telling at rates well above chance, though well below perfection.

Committee to Review the Scientific Evidence on the Polygraph, p. 4.

stimuli might be administered that made it impossible for a witness to do anything but tell the truth—an effective truth serum. (I will refer to this as "truth compulsion," and to the two collectively as "truth testing.") Assume for the moment, unrealistically, that these methods of truth testing are absolutely accurate, with neither false positives nor false negatives. How would, and should, courts treat the results of such truth testing? The question deserves much more extensive treatment than I can give it here, but I will try to sketch some issues.

Consider first the nonscientific issues of admissibility. One argument against admissibility was made by four justices of the Supreme Court in *United States v. Scheffer*,[26] a case involving a blanket ban on the admissibility of polygraph evidence. Scheffer, an enlisted man in the Air Force working with military police as an informant in drug investigations, wanted to introduce the results of a polygraph examination at his court-martial for illegal drug use.[27] The polygraph examination, performed by the military as a routine part of his work as an informant, showed that Scheffer denied illegal drug use during the same period that a urine test detected the presence of methamphetamine.[28] Military Rule of Evidence 707, promulgated by President George H. W. Bush in 1991, provides that "notwithstanding any other provision of law, the results of a polygraph examination, the opinion of a polygraph examiner, or any reference to an offer to take, failure to take, or taking of a polygraph examination, shall not be admitted into evidence."

The court-martial refused to admit Scheffer's evidence on the basis of Rule 707. His conviction was overturned by the Court of Appeals for the Armed Forces, which held that this per se exclusion of all polygraph evidence violated the Sixth Amendment.[29] The Supreme Court reversed in turn, upholding Rule 707, but in a fractured opinion. Justice Thomas wrote the opinion announcing the decision of the Court, which found

---

[26] 523 U.S. 303 (1998).

[27] Ibid., p. 305.

[28] Ibid., p. 306.

[29] 44 M.J. 442 (1996).

the rule constitutional on three grounds: continued question about the reliability of polygraph evidence, the need to "preserve the jury's core function of making credibility determinations in criminal trials," and the avoidance of collateral litigation.[30] Justices Rehnquist, Scalia, and Souter joined the Thomas opinion in full. Justice Kennedy, joined by Justices O'Connor, Ginsburg, and Breyer, concurred in the section of the Thomas opinion based on reliability of polygraph evidence. Those four justices did not agree with the other two grounds.[31] Justice Stevens dissented, finding that the reliability of polygraph testing was already sufficiently well established to invalidate any per se exclusion.[32]

Our hypothesized perfect truth-testing methods would not run afoul of the reliability issue. Nor, assuming the rules for admitted "truth-tested" evidence were sufficiently clear, would collateral litigation appear to be a major concern. Such testing would seem, however, even more than the polygraph, to evoke the concerns of the four justices about invading the sphere of the jury even when the witness had agreed to the use. Although at this point Justice Thomas's concern lacks the fifth vote it needs to become a binding precedent, the preservation of the jury's role might be seen by some courts as rising to a constitutional level under a federal or state constitutional right to a jury trial in criminal or civil cases. This issue could certainly be used as a policy argument against allowing such evidence, and, as an underlying concern of the judiciary, it might influence judicial findings under *Daubert* or *Frye* about the reliability of the methods.[33] Assuming robust proof of reliability, it is hard to see any other strong argument against the admission of this kind of evidence. (Whether Justice Thomas's rationale, either as a constitutional or a policy matter, would apply to nonjury trials seems more doubtful.)

On the other hand, some defendants might have strong arguments *for* the admission of such evidence, at least in criminal cases. Courts have found in the Sixth Amendment, perhaps in combination with

---

[30] 532 U.S. at 312–313.

[31] Ibid., p. 318.

[32] Ibid., p. 320.

[33] I owe this useful insight to Professor Fisher.

the Fifth Amendment, a constitutional right for criminal defendants to present evidence in their own defense. Scheffer made this very claim, that Rule 707, in the context of his case, violated his constitutional right to present a defense. The Supreme Court has two lines of cases dealing with this right. In *Chambers v. Mississippi*, the Court resolved the defendant's claim by balancing the importance of the evidence to the defendant's case with the reliability of the evidence.[34] In *Rock v. Arkansas*, a criminal defendant alleged that she could remember the events only after having her memory "hypnotically refreshed."[35] The Court struck down Arkansas's per se rule against hypnotically refreshed testimony on the ground that the rule, as a per se rule, was arbitrary and therefore violated the Sixth Amendment rights of a defendant to present a defense and to testify in her own defense. The *Rock* opinion also stressed that the Arkansas rule prevented the defendant from telling her own story in any meaningful way. That might argue in favor of the admissibility of a criminal defendant's own testimony, under truth compulsion, as opposed to an examiner giving his or her expert opinion about the truthfulness of the witness's statements based on the truth detector results. These constitutional arguments for the admission of such evidence would not seem to arise with the prosecution's case or with either the plaintiff's or the defendant's case in a civil matter (unless some state constitutional provisions were relevant).[36]

Assuming "truth-tested" testimony were admissible, should either a party's, or a witness's, offer or refusal to undergo truth testing be admissible in evidence as relevant to his or her honesty? Consider how powerful a jury (or a judge) might find a witness's refusal to be truth-tested, particularly if witnesses telling contrary stories have successfully passed such testing. Such a refusal could well prove fatal to the witness's credibility.

---

[34] 410 U.S. 284 (1973).

[35] 483 U.S. 44 (1987).

[36] A constitutional right to admit such evidence might also argue for a constitutional right for indigent defendants to have the government pay the cost of such truth testing, which might be small or great.

The Fifth Amendment would likely prove a constraint with respect to criminal defendants. The fact that a defendant has invoked the Fifth Amendment's privilege against self-incrimination cannot normally be admitted into evidence or considered by the trier of fact. Otherwise, the courts have held, the defendant would be penalized for having invoked the privilege. A defendant who takes the stand might well be held to have waived that right and so might be impeached by his refusal to undergo truth testing. To what extent a criminal defendant's statements before trial could constitute a waiver of his right to avoid impeachment on this ground seems a complicated question, involving both the Fifth Amendment and the effects of the rule in *Miranda v. Arizona*.[37] These complex issues would require a paper of their own; I will not discuss them further here.

Apart from a defendant in a criminal trial, it would seem that any other witnesses should be impeachable for their refusal to be truth-tested; they might invoke the privilege against self-incrimination, but the trier of fact, in weighing their credibility in this trial, would not be using that information against them. And this should be true for prosecution witnesses as well as defense witnesses. Both parties and nonparty witnesses at civil trials would seem generally to be impeachable for their refusal to be truth-tested, except in some jurisdictions that hold that a civil party's invocation of the Fifth Amendment may not be commented upon even in a civil trial.

It seems unlikely that a witness's *willingness* to undergo truth testing would add anything to the results of a test in most cases. It might, however, be relevant, and presumably admissible, if for some reason the test did not work on that witness or, unbeknownst to the witness at the time she made the offer, the test results turned out to be inadmissible.

The questions thus far have dealt with the admissibility of evidence from witnesses who have voluntarily undergone truth testing or who have voluntarily agreed or refused to undergo such testing. Could, or should, either side have the power to compel a witness to undergo either method of truth testing? At its simplest, this might be a right to retest a

---

[37] 396 U.S. 868 (1969).

witness tested by the other side, a claim that could be quite compelling if the results of these methods, like the results of polygraphy, were believed to be significantly affected by the means by which the test was administered—not just the scientific process but the substance and style of the questioning. More broadly, could either side compel a witness, in a criminal or a civil case, to undergo such truth testing either as part of a courtroom examination or in pretrial discovery?

Witnesses certainly can be compelled to testify, at trial or in deposition. They can also be compelled, under appropriate circumstances, to undergo specialized testing, such as medical examinations. (The latter procedures typically require express authorization from the court rather than being available as of right to the other side.) Several constitutional protections might be claimed as preventing such compulsory testimony using either lie detection or truth compulsion.

A witness might argue that the method of truth testing involved was so great an intrusion into the person's bodily (or mental) integrity as to "shock the conscience" and violate the Fifth or Fourteenth Amendment, as did the stomach pumping in *Rochin v. California*.[38] A test method involving something like the wearing of headphones might seem quite different from one involving an intravenous infusion of a drug or envelopment in the coffinlike confines of a full-sized MRI machine. The strength of such a claim might vary according to whether the process was lie detection and merely verified (or undercut) the witness's voluntarily chosen words, or whether it was truth compulsion and interfered with the witness's ability to choose her own words.

The Fifth Amendment's privilege against self-incrimination would usually protect those who choose to invoke it (and who have not been granted immunity). As noted above, that would not necessarily protect either a party in a civil case or a nondefendant witness in a criminal case from impeachment for invoking the privilege.

Would a witness have a possible Fourth Amendment claim that such testing, compelled by court order, was an unreasonable search and seizure by the government? I know of no precedent for considering questioning

---

[38] 342 U.S. 165 (1952).

itself as a search or seizure, but this form of questioning could be seen as close to searching the confines of the witness's mind. In that case, would a search warrant or other court order suffice to authorize the test against a Fourth Amendment claim? And, if it were seen in that light, could a search warrant be issued for the interrogation of a person under truth testing outside the context of any pending criminal or civil litigation—and possibly even outside the context of an arrest and Miranda rights that follow it? If this seems implausible, consider what an attractive addition statutory authorization of such "mental searches" might seem to the Bush administration or to Congress in the next version of the USA PATRIOT Act.[39]

In some circumstances, First Amendment claims might be plausible. Truth compulsion might be held to violate in some respects the right not to speak, although the precedents on this point are quite distant, involving a right not to be forced to say, or to publish, specific statements. It also seems conceivable that some religious groups could object to these practices and might be able to make a free exercise clause argument against such compelled speech.

These constitutional questions are many and knotty. Equally difficult is the question whether some or all of them might be held to be waived by witnesses who either had undergone truth testing themselves or had claimed their own truthfulness, thus "putting it in question." And, of course, even if parties or witnesses have no constitutional rights against being ordered to undergo truth testing, that does not resolve the policy issue of whether such rights should exist as a matter of statute, rule, or judicial decision.

Parties and witnesses are not the only relevant actors in trials. Truth testing might also be used in voir dire. Prospective jurors are routinely asked about their knowledge of the parties or of the case or about their relevant biases. Could a defendant claim that his right to an unbiased juror was infringed if such methods were not used and hence compel prospective jurors to undergo truth testing? Could one side or the other challenge

---

[39] Uniting and Strengthening America by Providing Appropriate Tools Required to Intercept and Obstruct Terrorism Act ("USA PATRIOT Act") of 2001, PL 107–156 (2001).

for cause a prospective juror who was unwilling to undergo such testing? In capital cases, jurors are asked whether they could vote to convict in light of a possible death penalty; truth testing might be demanded by the prosecution to make sure the prospective jurors are being honest.

It is also worth considering how the existence of such methods might change the pretrial maneuvers of the parties. Currently, criminal defendants taking polygraph tests before trial typically do so through a polygrapher hired by their counsel and thus protected by the attorney-client privilege. That may change. Whatever rules are adopted concerning the admissibility of evidence from truth testing will undoubtedly affect the incentives of the parties, in civil and criminal cases, to undergo truth testing. This may, in turn, have substantial, and perhaps unexpected, repercussions for the practices of criminal plea bargaining and civil settlement. As the vast majority of criminal and civil cases are resolved before trial, the effects of truth testing could be substantial.

Even more broadly, consider the possible effects of truth testing on judicial business generally. Certainly not every case depends on the honesty of witness testimony. Some hinge on conclusions about reasonableness or negligence; others are determined by questions of law. Even factual questions might be the focus of subjectively honest, but nevertheless contradictory, testimony from different witnesses. Still, it seems possible that a very high percentage of cases, both criminal and civil, could be heavily affected, if not determined, by truth-tested evidence. If truth testing reduced the number of criminal trials tenfold, that would surely raise Justice Thomas's concern about the proper role of the jury, whether or not that concern has constitutional implications. It would also have major effects on the workload of the judiciary and, perhaps, on the structure of the courts.

The questions raised by a perfect method of truth testing are numerous and complicated. They are also probably unrealistic, given that no test will be perfect. Most of these questions would require reconsideration if truth testing turned out to be only 99.9 percent accurate, or 99 percent accurate, or 90 percent accurate. That reconsideration would have to examine not just overall "accuracy" but the rates of both false positives (the identification of a false statement as true) and false negatives (the identification of a true statement as false), as those may have different implications. Similarly, decisions on admissibility might differ if accuracy rates varied with a witness's age, sex, training in "beating" the machine, or other traits. And, of course, proving the accuracy of such methods as they are first introduced or as they are altered will be a major issue in court systems under the *Daubert* or *Frye* tests.

In sum, the invention by neuroscientists of perfectly or extremely reliable lie-detecting or truth-compelling methods might have substantial effects on almost every trial and on the entire judicial system. How those effects would play out in light of our current criminal justice system, including the constitutional protections of the Bill of Rights, is not obvious.

DETERMINING BIAS

Evidence produced by neuroscience may play other significant roles in the courtroom. Consider the possibility of testing, through neuroimaging, whether a witness or a juror reacts negatively to particular groups. Already, neuroimaging work is going on that looks for—and finds— differences in the reaction of a subject's brain to people of different races. If that research is able to associate certain patterns of activity with negative bias, its possible use in litigation could be widespread.

As with truth testing, courts would have to decide whether bias testing met *Daubert* or *Frye*, whether voluntary test results would be admissible, whether a party's or witness's refusal or agreement to take the test could be admitted into evidence, and whether the testing could ever be compelled. The analysis on these points seems similar to that for truth testing, with the possible exception of a lesser role for the privilege against self-incrimination.

If allowed, neuroscience testing for racial bias might be used where bias was a relevant fact in the case, as in claims of employment discrimination based on race. It might be used to test any witness for bias for or against a party of a particular race. It might be used to test jurors to ensure that they were not biased against parties because of race. One could even, barely, imagine it being used to test judges for bias, perhaps as part of a motion to disqualify for bias. And, of course, such bias testing need not be limited to bias based on race, nationality, sex, or other protected groups. One could seek to test, in appropriate cases, for bias against parties or witnesses based on their occupation (the police, for example), their looks (too fat, too thin), their voices (a Southern accent, a Boston accent), or many other characteristics.

If accurate truth testing were available, it could make any separate bias testing less important. Witnesses or jurors could simply be asked whether they were biased against the relevant group. On the other hand, it is possible that people might be able to answer honestly that they were not biased, when they were in fact biased. Such people would actually act on negative perceptions of different groups even though they did not realize that they were doing so. If the neuroimaging technique were able to detect people with that unconscious bias accurately, it might still be useful in addition to truth testing.

Bias testing might even force us to reevaluate some truisms. We say that the parties to litigation are entitled to unbiased judges and juries, but we mean that they are entitled to judges and juries that are not demonstrably biased in a context where demonstrating bias is difficult. What if demonstrating bias becomes easy—and bias is ubiquitous? Imagine a trial in which neuroimaging shows that all the prospective jurors are prejudiced against a defendant who looks like a stereotypical Hell's Angel because they think he looks like a criminal. Or what if the only potential jurors who didn't show bias were themselves members of quasi-criminal motorcycle gangs? What would the defendant's right to a fair trial mean in that context?

EVALUATING OR ELICITING MEMORY

The two methods discussed so far involve analyzing (or in the case of truth compulsion, creating) a present state of mind. It is conceivable that neuroscience might also provide courts with at least three relevant tools concerning memory. In each case, courts would again confront questions of the reliability of the tools, their admissibility with the witness's permission, the impeaching of witnesses for failing to use the tools, and the compelling of a witness to use such a memory-enhancing tool.

The first tool might be an intervention, pharmacological or otherwise, that improved a witness's ability to remember events. It is certainly conceivable that researchers studying memory-linked diseases might create drugs that help people retrieve old memories or retrieve them in more detail. This kind of intervention would not be new in litigation. The courts have seen great controversy over the past few years over "repressed" or "recovered" memories, typically traumatic early childhood experiences brought back to adult witnesses by therapy or hypnosis. Similarly, some of the child sex abuse trials over the past decade have featured testimony from young children about their experiences. In both cases, the validity of these memories has been questioned. We do know from research that people will often come to remember, in good faith, things that did not happen, particularly when those memories have been suggested to them.[40] Similar problems might arise with "enhanced" memories.[41]

A second tool might be the power to assess the validity of a witness's memory. What if neuroscience could give us tools to distinguish between "true" and "false" memory? One could imagine different parts of a witness's brain being used while recounting a "true" memory, a "false" memory, or a creative fiction. Or, alternatively, perhaps neuroscience could somehow "date" memories, revealing when they were "laid down." These methods seem more speculative than either truth testing or bias testing, but if either one (or some other method of testing memory) turned out to be feasible, courts would, after the *Daubert* or *Frye* hearings, again face questions of admitting testimony concerning their voluntary use, allowing comment on a witness's refusal to take a test, and possibly compelling their use.

---

[40] As with bias detection, truth testing could limit the need for such memory assessment when the witness was conscious of the falsity of the memory. Memory assessment, however, could be useful in cases where the witness had actually come to believe in the accuracy of a questioned "false" memory.

[41] It is quite plausible that researchers might create drugs that help people make, retain, and retrieve new memories, important in conditions such as Alzheimer's disease. One can imagine giving such a drug in advance to someone whom you expected to witness an important event—although providing such a person with a video recorder might be an easier option.

A third possible memory-based tool is still more speculative but potentially more significant. There have long been reports that electrical stimulation can, sometimes, trigger a subject to have what appears to be an extremely detailed and vivid memory of a past scene, almost like reliving the experience. At this point, we do not know whether such an experience is truly a memory or is more akin to a hallucination; if it is a memory, how to reliably call it up; how many memories might potentially be recalled in this manner; or, perhaps most important, how to recall any specific memory. Whatever filing system the brain uses for memories seems to be, at this point, a mystery. Assume that it proves possible to cause a witness to recall a specific memory in its entirety, perhaps by localizing the site of the memory first through neuroimaging the witness while she calls up her existing memories of the event. A witness could then, perhaps, *relive* an event important to trial, either before trial or on the witness stand. One can even, just barely, imagine a technology that might be able to "read out" the witness's memories, intercepted as neuronal firings, and translate them directly into voice, text, or the equivalent of a movie for review by the finder of fact. Less speculatively, one can certainly imagine a drug that would improve a person's ability to retrieve specific long-term memories.

While a person's authentic memories, no matter how vividly they are recalled, may not be an accurate portrayal of what actually took place, they would be more compelling testimony than that provided by typically foggy recollections of past events. Once again, if the validity of these methods were established, the key questions would seem to be whether to allow the admission of evidence from such a recall experience, voluntarily undertaken; whether to admit the fact of a party's or witness's refusal or agreement to use such a method; and whether, under any circumstances, to compel the use of such a technique.[42]

---

[42] Although it is not relevant to judicial uses of the technology, note the possibility that any such memory recall method, if easily available to individuals in unsupervised settings, could be used, or abused, with significant consequences. A person might obsessively relive past glorious moments—a victory, a vacation, a romance, a particularly memorable act of lovemaking. A depressed person might dwell compulsively on bad memories. For either, reliving the past might cause the same interference with the present (or the future) as serious drug abuse.

OTHER LITIGATION-RELATED USES

Neuroscience covers a wide range of brain-related activities. The three areas sketched above are issues where neuroscience could conceivably have an impact on almost any litigation, but neuroscience might also affect any specific kind of litigation where brain function was relevant. Consider four examples.

The most expensive medical malpractice cases are generally so-called bad baby cases. In these cases, children are born with profound brain damage. Damages can be enormous, sometimes amounting to the cost of round-the-clock nursing care for 70 years. Evidence of causation, however, is often very unclear. The plaintiff parents allege that the defendants managed the delivery negligently, which led to a lack of oxygen that in turn caused the brain damage. Defendants, in addition to denying negligence, usually claim that the damage had some other, often unknown, cause. Jurors are left with a family facing a catastrophic situation and no strong evidence about what caused it. Trial verdicts, and settlements, can be extremely high, accounting in part for the high price of malpractice insurance for obstetricians. If neuroscience could reliably distinguish between brain damage caused by oxygen deprivation near birth and that caused earlier, these cases would have more accurate results, in terms of compensating families only when the damage was caused around the time of delivery. Similarly, if fetal neuroimaging could reveal serious brain damage before labor, those images could be evidence about the cause of the damage. (One can even imagine obstetricians insisting on prenatal brain scans before delivery in order to establish a baseline.) A more certain determination of causation should also lead to more settlements and less wasteful litigation. (Of course, in cases where neuroscience showed that the damage was consistent with lack of oxygen around delivery, the defendants' negligence would still be in question.)

In many personal injury cases, the existence of intractable pain may be an issue. In some of those cases there may be a question whether the plaintiff is exaggerating the extent of the pain. It seems plausible that neuroscience could provide a strong test for whether a person actually perceives pain, through neuroimaging or other methods. It might be able to show whether signals were being sent by the sensory nerves to the brain from the painful location on the plaintiff's body. Alternatively, it might locate a region of the brain that is always activated when a person feels pain or a pattern of brain activation that is always found during physically painful experiences. Again, by reducing uncertainty about a very subjective (and hence falsifiable) aspect of a case, neuroscience could improve the litigation system.

A person's competency is relevant in several legal settings, including disputed guardianships and competency to stand trial. Neuroscience might be able to establish some more objective measures that could be considered relevant to competency. (It might also reveal that what the law seems pleased to regard as a general, undifferentiated competency does not, in fact, exist.) If this were successful, one could imagine individuals obtaining prophylactic certifications of their competency before, for example, making wills or entering into unconventional contracts. The degree of mental ability is also relevant in capital punishment, where the Supreme Court has recently held that executing the mentally retarded violates the Eighth Amendment.[43] Neuroscience might supply better, or even determinative, evidence of mental retardation. Or, again, it may be that neuroscience would force the courts to recognize that "mental retardation" is not a discrete condition.

Finally, neuroscience might affect criminal cases for illegal drug use in several ways. Neuroscience might help determine whether a defendant was "truly" addicted to the drug in question, which could have some consequences for guilt and sentencing. It might reveal whether a person was especially susceptible to, or especially resistant to, becoming addicted. Or it could provide new ways to block addiction, or even pleasurable sensations, with possible consequences for sentencing and treatment. Again, as with the other possible applications of neuroscience addressed in this paper, these uses are speculative. It would be wrong to count on neuroscience to solve, *deus ex machina*, our drug problems. It does not seem irresponsible, however, to consider the possible implications of neuroscience breakthroughs in this area.[44]

---

[43] *Atkins v. Virginia*, 536 U.S. 304 (2003).

[44] At the same time, neuroscience could give rise to new drugs or drug equivalents. A neuroscience-devised trigger of pleasurable sensations—say, that would cause powerful orgasms—could function effectively as a powerful drug of abuse.

Confidentiality and Privacy

I am using these two often conflated terms to mean different things. I am using *confidentiality* to refer to the obligation of a professional or an entity to limit appropriately the availability of information about people (in this context, usually patients or research subjects). *Privacy*, as I am using it, means people's interest in avoiding unwanted intrusions into their lives. The first focuses on limiting the distribution of information appropriately gathered; the second concerns avoiding intrusions, including the inappropriate gathering of information. Neuroscience will raise challenges concerning both concepts.

Maintaining—and Breaking—Confidentiality

Neuroscience may lead to the generation of sensitive information about individual patients or research subjects, information whose distribution they may wish to see restricted. Personal health information is everywhere protected in the United States: by varying theories under state law, by new federal privacy regulations under the Health Insurance Portability and Accountability Act (HIPAA),[45] and by codes of professional ethics. Personal information about research subjects must also be appropriately protected under the Common Rule, the federal regulation governing most (but not all) biomedical research in the United States.[46] The special issue with neuroscience-derived information is whether some or all of it requires additional protection.

Because of concerns that some medical information is particularly dangerous to patients, physicians have sometimes kept separate medical charts detailing patients' mental illness, HIV status, or genetic diseases. Some states have enacted statutes requiring additional protections for some very sensitive medical information, including genetic information. Because neuroscience

---

[45] 45 C.F.R. §160; 101, et seq. (2003).

[46] Each federal agency's version of the Common Rule is codified separately, but see, for example, the version of the regulation adopted by the Department of Health and Human Services, 45 C.F.R. §§ 46.101–46.409 et seq. (2003).

information may reveal central aspects of a person's personality, cognitive abilities, and future, one could argue that it too requires special protection. Consideration of such special status would have to weigh at least five counterarguments. First, any additional record keeping or data protection requirements both increase costs and risk making important information unavailable to physicians or patients who need it. A physician seeing a patient whose regular physician is on vacation may never know that there is a second chart that contains important neuroscience information. Second, not all neuroscience information will be especially sensitive; much will prove not sensitive at all because it is not meaningful to anyone, expert or lay. Third, defining *neuroscience information* will prove difficult. In parallel area statutes defining genetic information have either employed an almost uselessly narrow definition (the results of DNA tests) or have opted for a wider definition encompassing all data about a person's genome. The latter, however, would end up including standard medical information that provides some data about a person's genetics: blood type, cholesterol level, skin color, and family history, and so on. Fourth, mandating special protection for a class of information sends the message that the information is especially important even if it is not. In genetics, it is argued that legislation based on such "genetic exceptionalism" increases a false and harmful public sense of "genetic determinism." Similar arguments might apply to neuroscience. Finally, given the many legitimate and often unpredictable needs for access to medical information, confidentiality provisions will often prove ineffective at keeping neuroscience information private, especially from the health insurers and employers who are paying for medical care. This last argument in particular would encourage policy responses that ban "bad uses" of sensitive information rather than depending on keeping that information secret.

Laws and policies on confidentiality also need to consider the limits on confidentiality. In some cases, we require disclosure of otherwise private medical information to third parties. Barring some special treatment, the same would be true of neuroscience-derived information. A physician (including, perhaps, a physician-researcher) may have an obligation to report to a county health agency or the Centers for Disease Control and Prevention neuroscience-derived information about a patient that is linked to a reportable disease (an MRI scan showing, for example, a case of new variant Creutzfeldt-Jakob disease, the human version of mad cow disease); to a motor vehicle department information linked to loss-of-consciousness disorders; and to a variety of governmental bodies information regarding a suspicion of child abuse, elder abuse, pesticide poisoning, or other topics as specified by statute. In some cases, it might be argued, as it has been in genetics, that a physician has a responsibility to disclose a patient's condition to a family member if the family member is at higher risk of the same condition as a result. Finally, neuroscience information showing an imminent and serious threat from a patient to a third party might have to be reported under the *Tarasoff* doctrine.[47] Discussion of the confidentiality of neuroscience-derived information needs to take all of these mandatory disclosure situations into account.

PRIVACY PROTECTIONS AGAINST MENTAL INTRUSIONS

Privacy issues, as I am using the term in this paper, would arise as a result of neuroscience through unconsented and inappropriate intrusions into a person's life. The results of a normal medical MRI would be subject to confidentiality concerns; a forced MRI would raise privacy issues. Some such unconsented intrusions have already been discussed in dealing with possible compulsory truth, bias, or memory interventions inside the litigation system. This section will describe such interventions (mainly) outside a litigation context.

Intrusions by the government are subject to the Constitution and its protections of privacy, contained in and emanating from the penumbra of the Bill of Rights. Whether or not interventions were permitted in the courtroom, under judicial supervision, the government might use them in other contexts, just as polygraphs are used in security clearance

---

[47] *Tarasoff v. Regents of University of California*, 17 Cal.#3d 425, 551 P. 2d 334, 131 Cal. Rptr. 14 (1976). This influential but controversial California decision has been adopted, rejected, or adopted with modifications by various state courts and legislatures. For a recent update, see Fillmore Buckner and Marvin Firestone, *Where the Public Peril Begins: 25 Years After Tarasoff J. Legal Med.* 21 (2000): 187.

investigations. All of these nonlitigation governmental uses share a greater possibility of abuse than the use of such a technology in a court-supervised setting.

Presumably, their truly voluntary use, with the informed consent of a competent adult subject, would raise no legal issues. Situations in which agreement to take the test could be viewed as less than wholly voluntary would raise their own set of sticky problems about the degree of coercion. Consider the possibility of truth tests for those seeking government jobs, benefits, or licenses. Admission to a state college (or eligibility for government-provided scholarships or government-guaranteed loans) might, for example, be conditioned on passing a lie detection examination on illegal drug use.

Frank compulsion might also be used, although it would raise constitutional questions under the Fourth and Fifth Amendments. One could imagine law enforcement officials deciding to interrogate one member of a criminal gang under truth compulsion in violation of Miranda and the Fifth Amendment (and hence to forgo bringing him to trial) to get information about his colleagues. Even if a person had been given a sufficiently broad grant of immunity to avoid any Fifth Amendment issues, would that really protect the interests of a person forced to undergo a truth compulsion process? Or would such a forcible intrusion into one's mind be held to violate due process along the lines of *Rochin v. California*?[48]

Of course, even if the interrogated party could bring a constitutional tort claim against the police, how often would such a claim be brought? And would we—or our courts—always find such interrogations improper? Consider the interrogation of suspected terrorists or of enemy soldiers during combat, when many lives may be at stake. (This also raises the interesting question of how the United States could protect its soldiers or agents from similar questioning.)

Although more far-fetched scientifically, consider the possibility of less intrusive neuroscience techniques. What if the government developed a neuroimaging device that could be used at a distance from a moving subject or one that could fit into the arch of an airport metal

---

detector? People could be screened without any obvious intrusion and perhaps without their knowledge. Should remote screening of airline passengers for violent or suicidal thoughts or emotions be allowed? Would it matter whether the airport had signs saying that all travelers, by their presence, consented to such screening?

Private parties have less ability than the government to compel someone to undergo a neuroscience intervention—at least without being liable to arrest for assault. Still, one can imagine situations in which private parties either frankly coerce or unduly influence someone else to undergo a neuroscience intervention. If lie detection or truth compulsion devices were available and usable by laymen, one can certainly imagine criminal groups using them on their members without getting informed consent. Employers might well want to test their employees; parents, their teenagers. If the intervention required a full-sized MRI machine, we would not worry much about private, inappropriate use. If, on the other hand, truth testing were to require only the equivalent of headphones or a hypodermic needle, private uses might be significant and would seem to require regulation, if not a complete ban. This would seem even more true if remote or unnoticeable methods were perfected.

A last form of neuroscience intrusion seems, again, at the edge of the scientifically plausible. Imagine an intervention that allowed an outsider to control the actions or motions, and possibly even the speech, emotions, or thoughts, of a person. Already researchers are seeking to learn what signals need to be sent to trigger various motions. Miguel Nicolelis of Duke University has been working to determine what neural activity triggers particular motions in rats and monkeys, and he hopes to be able to stimulate it artificially.[49] One goal is to trigger implanted electrodes and have the monkey's arm move in a predictable and controlled fashion. The potential benefits of this research are enormous,

---

[49] See M. A. L. Nicolelis, "Brain-Machine Interfaces to Restore Motor Function and Probe Neural Circuits." *Nature Reviews Neuroscience* 4 (2003): 417–22. For a broader discussion of Nicolelis's work, see Jose M. Carmena et al., "Learning to Control a Brain-Machine Interface for Reaching and Grasping by Primates," *Public Library of Science Biology*, vol. 1, issue 2 (November 2003), available at http://www.plosbiology. org/plosonline/?request=get-document&doi=10.1371/journal.pbio.0000042.

particularly to people with spinal cord injuries or other interruptions in their motor neurons. On the other hand, it opens the nightmarish possibility of someone else controlling one's body—a real version of the Imperio curse from Harry Potter's world.

Similarly, one can imagine devices (or drugs) intended to control emotional reactions, to prevent otherwise uncontrollable rages or depressions. One can imagine a court ordering implantation of such a device in sexual offenders to prevent the emotions that give rise to their crimes or, perhaps more plausibly, offering such treatment as an alternative to a long prison term. Surgical or chemical castration, an old-fashioned method of accomplishing a similar result, is already a possibility for convicted sex offenders in some states. Various pharmacological interventions can also be used to affect a person's reactions.

These kinds of interventions may never become more than the ravings of victims of paranoia, though it is at least interesting that the Defense Advanced Research Projects Administration (DARPA) is providing $26 million in support of Nicolelis's research through its Brain-Machine Interfaces program.[50] The use of such techniques on consenting competent patients could still raise ethical issues related to enhancement. Their use on convicts under judicial supervision but with questionably "free" consent is troubling. Their possible use on unconsenting victims is terrifying. If such technologies are developed, their regulation must be considered carefully.

Patents

Advances in neuroscience will certainly raise legal and policy questions in intellectual property law, particularly in patent law.[51] Fortunately, few of

---

[50] "DARPA to Support Development of Human Brain-Machine Interfaces," Duke University press release (August 15, 2002), found at http://www.dukenews.duke.edu/research/darpacontract5.html.

[51] I cannot think of any plausible issues in copyright or trademark law arising from neuroscience (except, of course, to the extent that litigation in either field might be affected by some of the possible methods discussed in the litigation section above). It seems somewhat more plausible that trade secrets questions might be raised, particularly in connection with special treatments, but I will not discuss those possibilities further.

those questions seem novel, as most seem likely to parallel issues already raised in genetics. In some important respects, however, the issues seem less likely to be charged than those encountered in genetics.

Two kinds of neuroscience patents seem likely. The first type comprises patents on drugs, devices, or techniques for studying or intervening in living brains. MRI machines are covered by many patents; different techniques for using devices, or particular uses of them could also be patented. So, for example, the first person to use an MRI machine to search for a particular atom or molecule might be able to patent that use, unless it were an obvious extension of existing practice. Similarly, someone using an MRI machine, or a drug, for the purpose of determining whether a subject was telling the truth could patent that use of the machine or drug, even if she did not own a patent on the machine or the drug itself.

The second type would be a patent on a particular pattern of activity in the brain. (I will refer to these as neural pattern patents.) The claims might be that a pattern could be used to diagnose conditions, or predict future conditions, or as an opportunity for an intervention. This would parallel the common approach to patenting genes for diagnosis, for prediction, and for possible gene therapy. Neuroimaging results seem the obvious candidates for this kind of patent, although the patented pattern might show up, for example, as a set of gene expression results revealed by microarrays or gene chips.

I will discuss the likely issues these kinds of patents raise in three categories: standard bioscience patent issues, "owning thoughts," and medical treatments.

Standard Bioscience Patent Issues

Patents in the biological sciences, especially those relating to genetics, have raised a number of concerns. Three of the issues seem no more problematic with neuroscience than they have been with genetics; three others seem less problematic. Whether this is troublesome, of course, depends largely on one's assessment of the current state of genetic patents. My own assessment is relatively sanguine; I believe we are muddling through the issues of genetic patents with research and treatment continuing to thrive. I am optimistic, therefore, that none of these standard patent issues will cause broad problems in neuroscience.

Two concerns involve patent monopoly. Some complain that patents allow the patent owner to restrict the use and increase the price of the patented invention, thus depriving some people of its benefits.[52] This is, of course, true of all patents and is a core idea behind the patent system: the time-limited monopoly provides the economic returns that encourage inventors to invent. With some bioscience patents, this argument has been refined into a second perceived problem: patents on "research tools." Control over a tool essential to the future of a particular field could, some say, give the patent owner too much power over the field and could end up retarding research progress. This issue has been discussed widely, most notably in the 1998 Report of the National Institutes of Health (NIH) Working Group on Research Tools, which made extensive recommendations on the subject.[53] Some neuroscience patents may raise concerns about monopolization of basic research tools, but it is not clear that those problems cannot be handled if and as they arise.

A third issue concerns the effects of patents on universities. Under the Bayh-Dole Act, passed in 1980, universities and other nonprofit organizations where inventions were made using federal grant or contract funds can claim ownership of the resulting inventions, subject to certain conditions. Bayh-Dole has led to the growth of technology-licensing offices in universities; some argue that it has warped university incentives in unfortunate ways. While neuroscience patents might expand the number of favored, money-making departments in universities, they seem unlikely to make a qualitative difference.

---

[52] Jon F. Merz, Antigone G. Kriss, Debra G.B. Leonard, and Mildred K. Cho, "Diagnostic Testing Fails the Test," *Nature* 415 (2002): 577–579.

[53] Report of the National Institutes of Health (NIH) Working Group on Research Tools (June 4, 1998), available at http://www.nih.gov/news/researchtools/.

Because neuroscience patents seem unlikely to pose the first three patent problems in any new or particularly severe ways does not mean those issues should be ignored. Individual neuroscience patents might cause substantial problems that call for intervention; the cumulative weight of neuroscience patents when added to other bioscience patents may make systemic reform of one kind or another more pressing. But the outlines of the problems are known.

Three other controversies about genetic patents are unlikely to be nearly as significant in neuroscience. They seem relevant, if at all, to neural pattern patents, not to device or process patents.

Two of the controversies grew out of patents on DNA sequences. In 1998, Rebecca Eisenberg and Michael Heller pointed out "the tragedy of the anticommons," the concern that having too many different patents for DNA sequences under different ownership could increase transaction costs so greatly as to foreclose useful products or research.[54] This issue was related to a controversy about the standards for granting patents on DNA sequences. Researchers were applying for tens of thousands of patents on small stretches of DNA without necessarily knowing what, if anything, the DNA did. Often these were "expressed sequence tags," or ESTs, stretches of DNA that were known to be in genes and hence to play some role in the body's function because they were found in transcribed form as messenger RNA in cells. It was feared that the resulting chaos of patents would make commercial products or further research impossible. This concern eventually led the Patent and Trademark Office to issue revised guidelines tightening the utility requirement for gene patents.

However strong or weak these concerns may be in genetics, neither issue seems likely to be very important in neuroscience (except of course in neurogenetics). There does not appear to be anything like a DNA sequence in neuroscience, a discrete entity or

---

54 Michael A. Heller and Rebecca S. Eisenberg, "Can Patents Deter Innovation? The Anticommons in Biomedical Research," *Science* 280 (May 1, 1998): 698–701, but see John P. Walsh, Ashish Arora, and Wesley M. Cohen, "Research Tool Patenting and Licensing and Biomedical Innovation," in *Patents in the Knowledge-Based Economy*, ed. W. M. Cohen and S. Merrill (National Academies Press, 2003), which finds no evidence for such a problem.

pattern that almost certainly has meaning, and potential scientific or commercial significance, even if that meaning is unknown. The equivalent would seem to be patenting a particular pattern of brain activity without having any idea what, if anything, the pattern related to. That was plausible in genetics because the sequence could be used as a marker for the still unknown gene; nothing seems equivalent in neuroscience. Similarly, it seems unlikely that hundreds or thousands of neural patterns, each patented by different entities, would need to be combined into one product or tool for commercial or research purposes.

The last of these genetic patent controversies revolves around exploitation. Some have argued that genetic patents have often stemmed from the alleged inventors' exploitation of individuals or indigenous peoples who provided access to or traditional knowledge about medicinal uses of living things, who had created and maintained various genetically varied strains of crops, or who had actually provided human DNA with which a valuable discovery was made. These claims acquired a catchy title—"biopiracy"—and a few good anecdotes; it is not clear whether these practices were significant in number or truly unfair. Neuroscience should face few if any such claims. The main patterns of the research will not involve seeking genetic variations from crops or other living things, nor does it seem likely (apart from neurogenetics) that searches for patterns found in unique individuals or distinct human populations will be common.

"Owning Thoughts"

Patents on human genes have been extremely controversial for a wide variety of reasons. Some have opposed them for religious reasons, others because they were thought not to involve true "inventions," others because they believed human genes should be "the common heritage of mankind," and still others because they believe such gene patents "commodify" humans. (Similar but slightly different arguments have raged over the patentability of other kinds of human biological materials, or of nonhuman life-forms.) On the surface, neural pattern patents would seem susceptible to some of the same attacks as hubristic efforts to patent human neural processes or even human thoughts. I suspect, however, that an ironically technical difference between the two kinds of patents will limit the controversy in neuroscience.

Patents on human genes—or, more accurately, patents on DNA or RNA molecules of specified nucleotide sequences—are typically written to claim a wide range of conceivable use of those sequences. A gene patent, for example, might claim the use of a sequence to predict, to diagnose, or to treat a disease. But it will also claim the molecule itself as a "composition of matter." The composition-of-matter claim gives the owner rights over any other uses of the sequence, even though he has not foreseen them. It also seems to give him credit for "inventing" a genetic sequence that exists naturally and that he merely isolated and identified. It is the composition-of-matter claims that have driven the controversy over gene patents. Few opponents claim that the researchers who, for example, discovered the gene linked to cystic fibrosis should not be able to patent beneficial uses of that gene, such as in diagnosis or treatment. It is the assertion of ownership of the thing itself that rankles, even though that claim may add little value to the other "use" claims.

Neural pattern patents would differ from gene patents in that there is no composition of matter to be patented. The claim would be to certain patterns *used* for certain purposes. The pattern itself is not material—it is not a structure or a molecule—and so should not be claimable as a composition of matter. Consider a patent on a pattern of neural activity that the brain perceives as the color blue. A researcher might patent the use of the pattern to tell if someone was seeing blue, or perhaps to allow a person whose retina did not perceive blue to "see" blue. I cannot see how a patent could be issued on the pattern itself such that a person would "own" the "idea of blue." Similarly, a pattern that was determinative of schizophrenia could be patented for that use, but the patentee would not "own" schizophrenia or even the pattern that determined it. If a researcher created a pattern by altering cells, then he could patent, as a composition of matter, the altered cells, perhaps defined in part by the pattern they created. Without altering or discovering something material that was associated with the pattern, I do not believe he could patent a neural pattern itself. The fact that neural pattern patents will be patents to uses of the patterns, not for the patterns themselves, may well prevent the kinds of controversies that have attended gene patents.

PATENTS AND MEDICAL TREATMENT

Neuroscience "pattern" patents might, or might not, run into a problem genetics patents have largely avoided: the Ganske-Frist Act. In September 1996, as part of an omnibus appropriations bill, Congress added by amendment a new section, 287(c), to the patent law. This section states that

> with respect to a medical practitioner's performance of a medical activity that constitutes an infringement under section 271(a) or (b) of this title, the provisions of sections 281, 283, 284, and 285 of this title shall not apply against the medical practitioner or against a related health care entity with respect to such medical activity.[55]

This section exempts a physician and her hospital, clinic, HMO, or other "related health care entity" from liability for damages or an injunction for infringing a patent during the performance of a "medical activity." The amendment defines "medical activity" as "the performance of a medical or surgical procedure on a body," but it excludes from that definition "[1] the use of a patented machine, manufacture, or composition of matter in violation of such patent, [2] the practice of a patented use of a composition of matter in violation of such patent, or [3] the practice of a process in violation of a biotechnology patent."[56] The statute does not define "a biotechnology patent."

Congress passed the amendment in reaction to an ultimately unsuccessful lawsuit brought by an ophthalmologist who claimed that

---

[55] 35 U.S.C. 287(c) (2003).

[56] 35 U.S.C. 287(c) (2) (a) (2003).

another ophthalmologist infringed his patent on performing eye surgery using a particular V-shaped incision. Medical procedure patents had been banned in many other countries and had been controversial in the United States for over a century; they had, however, clearly been allowed in the United States since 1954.[57]

Consider a neural pattern patent that claimed the use of a particular pattern of brain activity in the diagnosis or as a guide to the treatment of schizophrenia.[58] A physician using that pattern without permission would not be using "a patented machine, manufacture, or composition of matter in violation of such patent." Nor would she be engaged in "the practice of a patented use of a composition of matter in violation of such patent." With no statutory definition, relevant legislative history, or judicial interpretation, it seems impossible to tell whether she would be engaged in the "practice of a process in violation of a biotechnology patent." Because molecules, including DNA, RNA, and proteins, can be the subjects of composition-of-matter patents, most genetic patents should not be affected by the Ganske-Frist Act.[59] Neural pattern patents might be. It is, of course, quite unclear how significant an influence this exception for patent liability might have in neuroscience research or related medical practice.

Conclusion

If even a few of the possibilities discussed above come to pass, neuroscience will have broad effects on our society and our legal system. The project to which this paper contributes can help in beginning to sift

---

[57] See the discussion of the Ganske-Frist amendment in Richard P. Burgoon, Jr., *Silk Purses, Sows Ears and Other Nuances Regarding 35 U.S.C. §287(c)*, U. Balt. Intell. Prop. J. 4 (1996): 69, and Scott D. Anderson, *A Right Without a Remedy: The Unenforceable Medical Procedure Patent*, Marq. Intell. Prop. L. Rev. 3 (1999): 117.

[58] If the use were purely for prediction, it could plausibly be argued that it was not a "medical procedure" subject to the act. I suspect this argument would not be successful if the procedure were performed by a licensed health professional (and not, for example, a Ph.D. neuroscientist).

[59] Procedures using gene expression results might be vulnerable unless the expression array or gene chip was itself a patented machine or manufacture the use of which was specified in the patent.

out the likely from the merely plausible, the unlikely, and the bizarre, both in the expected development of the science and in the social and legal consequences of that science. Truly effective prediction of upcoming problems—and suggestions for viable solutions—will require an extensive continuing effort. How to create a useful process for managing the social and legal challenges of neuroscience is not the least important of the many questions raised by neuroscience.

New Neuroscience, Old Problems

STEPHEN J. MORSE, J.D., PH.D.[1]

*Ferdinand Wakeman Hubbell Professor of Law and Professor of
Psychology and Law in Psychiatry, University of Pennsylvania.*

*If you want to know how the mind works, you should
investigate the mind, not the brain, and still less the
genome . . . It is clear that genes build proteins, but God
only knows what happens next . . .*[2]

Introduction

This essay considers a number of legal, moral, and political issues
that appear to be raised by increasingly sophisticated neuroscientific
understanding of the brain, nervous system, and behavior. Central
topics addressed include free will and responsibility, the desirability and
permissibility of enhancements of cognitive abilities that are within the
normal range, informed consent, and the development of legal doctrine

[1] This paper was presented at a workshop on Neuroscience and the Law sponsored
by the American Association for the Advancement of Science and the Dana
Foundation. I thank Ed Greenlee for his help, and, as always, I thank my personal
attorney, Jean Avnet Morse, for her sound, sober counsel and moral support.
The participants at the workshop made many helpful comments that have been
incorporated in the final version of this paper.

[2] Jerry Fodor, "Crossed Wires." *Times Literary Supplement* 3 (May 15, 2003).

generally. It will also briefly examine the admissibility of new scientific evidence in trials that raise issues for which neuroscientific evidence may be relevant.

The essay's central theme is that the new neuroscience in its current state may have fewer deep normative implications for law and society than popular imagination and even many scientists believe. Moreover, there are few observable trends in current law that reflect neuroscientific discoveries, which is not surprising or objectionable if the central claim about limited implications is justified.

The argument is conceptual, as it must be, and not primarily empirical. To address the relevance or implications of anything for law, it is important to have some understanding of what law is. Many take for granted that they know the answer to this question, but it is philosophically fraught. Nevertheless, a working understanding of what law is is necessary, and this paper provides a simplistic but helpful model. Then I turn to the law's concept of the person because law is addressed to and concerns persons and because the new neuroscience potentially alters our understanding of persons. Next, I consider the law's theory of individual responsibility and competence generally. With an understanding of law and the law's concept of the person and responsibility in place, we will then be in a position to consider the legal implications of neuroscience. I pay special attention to the alleged problem of free will, because potentially undermining the concept of free will appears to be the most profound challenge to the law's foundational view of the person.

The Law's Concept of the Person

Law is a socially constructed, intensely practical and evaluative system of rules and institutions that guides and governs human action, that helps us live together. It tells citizens what they may, must, and may not do, and what they are entitled to, and it includes institutions to ensure that law is made and enforced. Although questions such as whether there are necessary and sufficient conditions for something to be a law and what gives law its authority are intensely controversial, virtually

all commentators would agree with the foregoing description. Law
gives people good reason to behave one way or another by making the
consequences of noncompliance clear or through people's understanding
of the reasons that support a particular rule. As an action-guiding system
of rules, law shares many characteristics with other sources of guidance,
such as morality and custom, but law is distinguishable because its rules
and institutions are created and enforced by the state.

The law's concept of the person flows from the nature of law itself
and may be usefully contrasted for our purposes with a more mechanistic
view of the person. To introduce the issues, allow me to describe a
demonstration with which I often begin a presentation to neuroscientists
and psychologists. I ask the audience to assist me with a demonstration
that is loosely based on Wittgenstein's famous question, "What is left
over if I subtract the fact that my arm goes up from the fact that I raise
my arm?"[3] I vary the instructions somewhat, but the modal request is
to ask people who have brown hair and who wear eyeglasses to raise
their nondominant arm. Although I have never rigorously observed and
calculated the rate of cooperation, eyeballing my audiences suggests that
I get total cooperation. No one who believes that he or she has brown
hair (the instructions clarify that the current color is brown), wears
eyeglasses (the instructions clarify that anyone who wears eyeglasses
at any time is included), and is able to identify his or her nondominant
arm (I assume that everyone in a professional audience can perform this
task) fails to raise that arm. I then tell the participants that they may
lower their arms, and I politely thank them for their cooperation. Next
I ask the participants to explain what caused their arms to rise. I usually
then paraphrase this question as, "Why did you raise your arms?" How
should one answer such a question?

Bodily movements that appear to be the results of intentions—that
is, *actions*—unlike other phenomena, can be explained by physical causes
and by reasons for action. This is also true of intentionally produced
mental states or feelings, although we do not tend to term the production

[3] Ludwig Wittgenstein, *Philosophical Investigations*. 3d ed., part I, section 621, trans.
G. E. M. Anscombe, New York: Macmillan, 1958.

of such states actions. Although physical causes explain the structure and
mechanisms of the brain and nervous system and all the other moving
parts of the physical universe, only human action and other intentionally
produced states can also be explained by reasons. When one asks about
human action, "Why did she do that?" two distinct types of answers may
therefore be given. The reason-giving explanation accounts for human
behavior as a product of intentions that arise from the desires and beliefs
of the agent. The second type of explanation treats human behavior as
simply one more bit of the phenomena of the universe, subject to the
same natural, physical laws that explain all phenomena.

Human action is distinguished from all other phenomena because
only action is explained by reasons resulting from desires and beliefs,
rather than simply by mechanistic causes.[4] Only human beings are fully
intentional creatures. To ask why a person acted a certain way is to ask
for reasons for action, not for reductionist biophysical, psychological,
or sociological explanations. To comprehend fully why an agent has
particular desires, beliefs, and reasons requires biophysical, psychological,
and sociological explanations, but ultimately, human action is not
simply the mechanistic outcome of mechanistic variables.[5] Only persons
can deliberate about what action to perform and can determine their
conduct by practical reason. Suppose, for example, we wish to explain
why the participants raised their nondominant arms. The reason-giving
explanation might be that they desired the presentation to go well and
believed that if they cooperated, they would increase the probability that
this would happen. Therefore, they formed the intention to raise their
nondominant arms and executed that intention. This type of causal

---

[4] I recognize that there is no uncontroversial definition of action in any of the relevant
literatures. See, for example, Robert Audi, *Action, Intention and Reason* 1–4 (Ithaca:
Cornell University Press, 1993) (describing the basic philosophical divisions in each
of the four major problem areas in action theory). Nonetheless, the account I am
giving is most consistent with the law's view, as is explained below.

[5] I assume, controversially to be sure, that mental states are not identical to brain
states, even though the former depend on the latter. In the argot of philosophy of
mind, I am assuming that material reductionism is not the best explanation of the
relation between mental states and brain states. See pp. 163–168. My assumption is
consistent with the law's view of the person.

explanation, which philosophers and some psychologists refer to as folk psychology, is clearly the dominant type of explanation we all use every day to understand ourselves, other people, and our interactions with others in the world.

The mechanistic type of explanation would approach these questions quite differently. For example, those who believe that the mind can ultimately be reduced to the biophysical workings of the brain and nervous system or that mental states have no ontological status whatsoever would believe that the participants' arm raisings are *solely* the law-governed product of biophysical causes. Desires, beliefs, intentions, and choices, if they exist at all, are therefore simply correlational or epiphenomenal, rather than genuine causes of behavior. According to this mode of explanation, human behavior is indistinguishable from any other phenomena in the universe, including the movements of molecules and bacteria. It is always a bit amusing and disconcerting to observe neuroscientists struggling to explain the arm raisings in the language of mechanisms, as they sometimes do, especially if I push them, because they so clearly do not have a real clue about how it happens. I am not implying the unduly skeptical claim that we have no neurological and musculoskeletal understanding of the necessary conditions for bodily movements. It is simply true, however, that despite the extraordinary progress in neuroscience, how human action occurs—the so-called mind-body problem—remains fundamentally mysterious.

The social sciences, including psychology and psychiatry, are uncomfortably wedged between the reason-giving and the mechanistic accounts of human behavior. Sometimes they treat behavior "objectively," as primarily mechanistic or physical, sometimes "subjectively," as a text to be interpreted. Other times, social science engages in an uneasy amalgam of the two. What is always clear, however, is that the domain of the social sciences is human *action* and not simply the movements of bodies in space. One can attempt to assimilate folk psychology's reason giving to mechanistic explanation by claiming that desires, beliefs, and intentions are genuine causes, and not simply rationalizations of behavior. Indeed, folk psychology proceeds on the assumption that reasons for action are genuinely causal. But the assimilationist position is philosophically controversial, a controversy that will not be resolved until the mind-body problem is "solved."

At present, however, we have no idea how the brain enables the mind, but when we solve this problem—if we ever do—the solution will revolutionize our understanding of biological processes.[6] On the other hand, some philosophers deny that human beings have the resources ever to solve the mind-body problem and to explain consciousness, which is perhaps the greatest mystery.[7] Assuming that these problems can be unraveled, our view of ourselves and all our moral and political arrangements are likely to be as profoundly altered as our understanding of biological processes. Again, however, despite the impressive gains in neuroscience and related disciplines, we still do not know mechanistically how action happens, even if we are convinced, as I am, that a physicalist account of some sort must be correct.

Law, unlike mechanistic explanation or the conflicted stance of the social sciences, views human action as reason-governed and treats people as intentional agents, not simply as part of the biophysical flotsam and jetsam of the causal universe. It could not be otherwise, because law is concerned with human action. It makes no sense to ask a bull that gores a matador, "Why did you do that?", but this question makes sense and is vitally important when it is addressed to a person who sticks a knife into the chest of another human being. It makes a great difference to us if the knife wielder is a surgeon who is cutting with the patient's consent, or a person who is enraged at the victim and intends to kill him.

Law is a system of rules that at the least is meant to guide or influence behavior and thus to operate as a potential cause of behavior. As John Searle writes,

> Once we have the possibility of explaining particular forms of human behavior as following rules, we have a very rich explanatory apparatus

---

6   See Paul R. McHugh and Philip R. Slavney, *The Perspectives of Psychiatry*, 2d ed. (Baltimore: Johns Hopkins University Press, 1998) pp. 11–12.

7   See for example, Colin McGinn, *The Mysterious Flame: Conscious Minds in a Material World* (New York: Basic Books, 1999).

that differs dramatically from the explanatory apparatus of the natural sciences. When we say we are following rules, we are accepting the notion of mental causation and the attendant notions of rationality and existence of norms . . . The content of the rule does not just describe what is happening, but plays a part in *making it happen.*[8]

But legal and moral rules are not simply mechanistic causes that produce "reflex" compliance. They operate within the domain of practical reason. People are meant to and can only use these rules as potential *reasons for action* as they deliberate about what they should do. Moral and legal rules are thus action guiding primarily because they provide an agent with good moral or prudential reasons for forbearance or action. If people were not capable of understanding and then using legal rules as premises in deliberation, law would be powerless to affect human behavior.[9] People use legal rules as premises in the practical syllogisms that guide much human action. No "instinct" governs how fast a person drives on the open highway. But among the various explanatory variables, the posted speed limit and the belief in the probability of paying the consequences for exceeding it surely play a large role in the driver's choice of speed. I am not suggesting that human behavior cannot be modified by means other than influencing deliberation or that human beings always deliberate before they act. Of course it can, and of course they don't. But law operates through practical reason, even when we most habitually follow the legal rules. Law can directly and indirectly affect the world we inhabit only by its influence on practical reason.

---

[8] John R. Searle, "End of the Revolution," *New York Review of Books*, February 28, 2002, pp. 33, 35.

[9] See Scott J. Shapiro, *Law, Morality and the Guidance of Conduct*, 6 *Legal Theory* 127, 131 (2000). This view assumes that law is sufficiently knowable to guide conduct, but a contrary assumption is largely incoherent. As Shapiro writes,

> Legal skepticism is an absurd doctrine. It is absurd because the law cannot be the sort of thing that is unknowable. If a system of norms were unknowable, then that system would not be a legal system. One important reason why the law must be knowable is that its function is to guide conduct.

> I do not assume that legal rules are always clear and thus capable of precise action guidance. If most rules in a legal system were not sufficiently clear most of the time, however, the system could not function.

For the law, then, a person is a practical reasoner. The legal view is not that all people always reason and behave consistently rationally according to some preordained, normative notion of rationality. It is simply that people are creatures who are capable of acting for and consistently with their reasons for action and who are generally capable of minimal rationality according to mostly conventional, socially constructed standards of rationality.

A perfectly plausible evolutionary story explains why human beings need rules such as those law provides. We have evolved to be self-conscious creatures who act for reasons. Practical reason is inescapable for creatures like ourselves who inevitably care about the ends they pursue and about what reason they have to act in one way rather than another.[10] Because we are social creatures whose interactions are not governed primarily by innate repertoires, it is inevitable that rules will be necessary to help order our interactions in any minimally complex social group.[11] Human beings have developed extraordinarily diverse ways of living together, but the ubiquitous feature of all societies is that they are governed by rules addressed to beings capable of following those rules. The most basic view of human nature is that we are rational creatures. As I shall discuss below, the new neuroscience does not yet pose a threat to this fundamental conception and all that follows from it, including the concept of responsibility, to which this essay now turns.

The Legal Concept of Responsibility

The law's concept of responsibility follows logically from its conception of the person and the nature of law itself. As we have seen, law can guide action only if human beings are rational creatures who can understand and conform to legal requirements through intentional action. Legally

---

[10] Hilary Bok, *Freedom and Responsibility* (Princeton: Princeton University Press, 1998), pp. 75–91, 129–131, 146–151.

[11] Larry Alexander and Emily Sherwin, *The Rule of Rules: Morality, Rules & the Dilemmas of Law* (Durham: Duke University Press, 2001) pp. 11–25 (explaining why rules are necessary in a complex society and contrasting their account with H. L. A. Hart's theory).

responsible or legally competent agents are therefore people who have the general capacity to grasp and to be guided by good reason in particular legal contexts, who are generally capable of properly using the rules as premises in practical reasoning. As I shall discuss in the next two sections, responsibility, properly understood, has nothing to do with what most people understand by "free will." Rationality is the touchstone of responsibility. What rationality demands will of course differ across contexts. The requirements for competence to contract and for criminal responsibility are not identical. The usual legal presumptions are that adults are capable of minimal rationality and responsibility and that the same rules may be applied to all.

The law's requirement for responsibility—a general capacity for rationality—is not self-defining. It must be understood according to some contingent, normative notion both of rationality and of how much capability is required. For example, legal responsibility might require the capability of understanding the reason for an applicable rule, as well as the rule's narrow behavior command. These are matters of moral, political, and, ultimately, legal judgment, about which reasonable people can and do differ. There is no uncontroversial definition of rationality or of what kind and how much is required for responsibility in various legal contexts. These are normative issues, and, whatever the outcome might be within a polity and its legal system, the debate is about human action—intentional behavior that is potentially rationalized and guidable by reasons.

The Neuroscience Challenge to Personhood and Responsibility

The seriousness of neuroscience's potential challenge to the traditional legal concepts of personhood and responsibility is best summed up in the title of an eminent psychologist's recent book—*The Illusion of Conscious Will,* by Daniel Wegner.[12] In brief, the thesis is that we delude ourselves when we think that our intentions are genuinely causal. In fact, we are

---

[12]   Daniel M. Wegner, *The Illusion of Conscious Will* (Cambridge: MIT Press, 2002). Wegner is a professor of psychology at Harvard University.

just mechanisms, although the illusion of "conscious will" may play a positive role in our lives. Wegner's evidence and arguments are not all based on neuroscientific findings, but the claim that we are purely mechanisms is often thought to follow from all physicalist, naturalist, and scientific views of the person. If this is right, our conception of human nature must change, and the foundations of law and morality appear insupportable. This ominous prospect is not imminent, however. Advances in neuroscience and related fields have revealed hitherto unimagined biological causes that predispose people to behave as they do,[13] but the science typically supporting claims that conscious will is an illusion—that we do not act and are not responsible—either is insufficient empirically to support such a claim or does not have the implications supposed. In this section, I begin with the challenge to personhood and then turn to the challenge to responsibility.

THE CHALLENGE TO PERSONHOOD AND ACTION

The philosophy of mind and action has long contained arguments for various forms of material reductionism and for eliminative

---

[13] Recently published issues of prestigious journals contain illustrative examples of advances in scientific understanding of the causes of legally relevant behavior. See, for example, Avshalom Caspi et al., "Role of Genotype in the Cycle of Violence in Maltreated Children," 297 *Science* 851 (2002). (maltreated male children were more likely to exhibit antisocial behavior if they had a defect in the genotype that confers high levels of the neurotransmitter-encoding enzyme, monoamine oxidase A, which metabolizes various neurotransmitters linked to violence if the levels of those neurotransmitters are low); Rita Z. Goldstein and Nora D. Volkow, "Drug Addiction and Its Underlying Neurobiological Basis: Neuroimaging Evidence for the Involvement of the Frontal Cortex," 159 *American Journal of Psychiatry* 1542 (2002) (addiction involves cortical processes that result in the overvaluation of drug reinforcers, the undervaluation of other reinforcers, and defective inhibitory control of responses to drugs); Marc N. Potenza et al., "Gambling Urges in Pathological Gambling: A Functional Magnetic Resonance Imaging Study," 60 *Archives of General Psychiatry* 828 (2003) (when viewing gambling cues, male pathological gamblers "demonstrate relatively decreased activity in brain regions implicated in impulse regulation compared with controls"); Murray B. Stein et al., "Genetic and Environmental Influences on Trauma Exposure and Posttraumatic Stress Disorder Symptoms: A Twin Study," 159 *American Journal of Psychiatry* 1675 (2002) (concluding that genetic factors can influence the risk of exposure to assaultive trauma and to PTSD symptoms that may ensue).

materialism.[14] Reductive accounts hold, simply, that mental states are as they seem to us, but that they are identical to brain states. Eliminative accounts hold that our beliefs about our mental states are radically false and, consequently, that no matchup between brain states and mental states is possible. Both types of views are conceptual and existed long before the exciting recent discoveries in neuroscience and psychology that have so deepened our understanding of how the brain and nervous system are constructed and work. Needless to say, both are extremely controversial. Most philosophers of mind believe that complete reduction of mind to biophysical explanation is impossible.[15] Again, until the conceptual revolution that allows us to solve the mind-body problem occurs, science cannot resolve the debate, although it can furnish support for conceptual arguments. At present and for the foreseeable future, no one can demonstrate irrefutably that we are "merely" ultracomplicated biophysical machines. We have no convincing conceptual reason from the philosophy of mind, even when it is completely informed about the most recent neuroscience, to abandon our view of ourselves as creatures with causally efficacious mental states.

Even if we cannot solve the mind-body problem, however, and thus determine if reductive accounts are true, it is possible that we might make empirical discoveries indicating that some parts of our ordinary understanding about action and agency are incorrect. Much recent argument based on current neuroscience and psychology takes this position, arguing that mental causation does not exist as we think it does. For ease of exposition, let us call this the "no action thesis," or NAT. But the logic of these arguments is often shaky. Discovering a brain correlate or cause of an action does not mean that it is not an action. If actions exist, they have causes, including those arising in the brain.

---

[14] See Paul M. Churchland, *Matter and Consciousness*, rev. ed., pp. 26–34, 43–49 (Cambridge: MIT Press, 1988) (explaining the arguments for and against both types of arguments).

[15] See, for example, Galen Strawson, "Consciousness, Free Will, and the Unimportance of Determinism," 32 *Inquiry* 3 (1989) (claiming that reductive physicalism about the mind is "moonshine"); see, generally, John R. Searle, *The Rediscovery of the Mind* (Cambridge: MIT Press, 1992) (providing an extended argument for the irreducible reality of mind).

The real question is whether scientific, empirical studies have shown that action is rare or nonexistent, that conscious will is an illusion after all. Two kinds of evidence are often adduced: first, demonstrations that a very large part of our activity is undeniably caused by variables we are not in the slightest aware of, and, second, studies indicating that more activity than we think takes place when our consciousness is divided or diminished. Neither kind of evidence offers logical support to NAT, however. Just because a person may not be aware of the "real" causes or of all the causes for why she formed an intention does not entail that she did not form an intention and was not a fully conscious agent when she did so. Even if human beings were *never* aware of the causes of their intentions to act and their actions, it still would not mean that they were not acting intentionally and consciously.

Human consciousness can undeniably be divided or diminished by a wide variety of normal and abnormal causes.[16] We knew this long before contemporary scientific discoveries of what causes such states and how they correlate with brain structure and processes. Law and morality agree that if an agent's capacity for consciousness is nonculpably diminished, responsibility is likewise diminished. Some believe that it is diminished because bodily movements in the absence of consciousness are not actions.[17] Others believe that apparently goal-directed behavior that is responsive to the environment, such as sleepwalking, is action, but that it should be excused because diminished consciousness reduces the capacity for rationality.[18] Let us assume that the former view is correct, however, because it offers more direct moral and legal support to NAT. Let us also assume that studies have demonstrated that divided or diminished consciousness is more common than we think.

---

[16] See Jeffrey L. Cummings and Michael S. Mega, *Neurospsychiatry and Behavioral Neuroscience*, pp. 333–43 (Oxford: Oxford University Press, 2003) (description of dissociative and related states and their causes and treatments).

[17] For example, Michael S. Moore, *Act and Crime* 49–52, 135–155, 257–58 (1993); *More on Act and Crime*, 142 *University of Pennsylvania LR* 1749, 1804–20 (1994). The *Model Penal Code* takes this position, sec. 2.01.

[18] For example, Stephen J. Morse, *Culpability and Control*, 142 University of Pennsylvania LR 1587, 1641–52; Bernard Williams, *The Actus Reus of Dr. Caligari*, 142 *University of Pennsylvania LR* 1661 (1994).

To demonstrate that divided or partial consciousness is more common than it appears certainly extends the range of cases in which people are not responsible or have diminished responsibility, but such studies do not demonstrate that most human bodily movements that appear intentional, that appear to be actions, do not occur when the person has reasonably integrated consciousness. One cannot generalize from deviant cases or cases in which a known abnormality is present.

What is needed to support NAT is thus a general demonstration that causal intentionality is an illusion *tout court*, but I believe that no such general demonstration has yet been produced by scientific study. The most interesting evidence has arisen from studies done by Benjamin Libet,[19] which have generated an immense amount of comment.[20] Briefly, Libet's exceptionally creative and careful studies demonstrate that measurable electrical brain activity associated with intentional actions occurs about 550 milliseconds before the subject actually acts, and for about 350–400 milliseconds *before* the subject is consciously aware of the intention to act. Let us assume the validity of the studies.[21] I do not think they imply, however, that conscious intentionality does no causal work. They simply demonstrate that nonconscious brain events precede conscious experience, but this seems precisely what one would expect of the mind-brain. It does not mean that intentionality plays no causal role, and Libet concedes that people can "veto" the act, which is another form of mental act that plays a causal role. Libet's work is fascinating, but it does not prove that persons are not conscious, intentional agents.

---

[19] Benjamin Libet, "Do We Have Free Will?" in Benjamin Libet, Anthony Freeman, and Keith Sutherland, eds., *The Volitional Brain: Towards a Neuroscience of Free Will* 47 (Thorverton, UK: Imprint Academic, 1999) (summarizing the findings and speculating about their implications).

[20] Wegner, see note 12, pp. 54–55 (characterizing the recounting of Libet's results as a "cottage industry" and noting the large and contentious body of commentary).

[21] For a more cautious critique of Libet's investigations, see Jing Zhu, "Reclaiming Volition: An Alternative Interpretation of Libet's Experiment," 10 J. *Consciousness Studies* 61, 67–74 (2003) (providing an alternative interpretation of Libet's results based on artifacts in Libet's experimental design and on the results of experiments by other investigators).

NAT provides no guidance about what we should do next and is
potentially incoherent. Let us suppose that you were convinced by the
mechanistic view of persons that you were not an intentional, rational
agent after all. (Of course, the notion of being "convinced" would be
an illusion, too. Being convinced means that you were persuaded by
evidence or argument, but a mechanism is not persuaded by anything. It
is simply neurophysically transformed or some such.) What should you
do now? You know that it's an illusion to think that your deliberation
and intention have any causal efficacy in the world. (Again, what
does it mean according to the purely mechanistic view to "know"
something? But enough.) You also know, however, that you experience
sensations such as pleasure and pain and that you care about what
happens to you and to the world. You cannot just sit quietly and wait
for your neurotransmitters to fire. You will, of course, deliberate and
act. Even if pure mechanism is true—about which, once again, we
will never be certain until we solve the mind-body problem—human
beings will find it almost impossible not to treat themselves as rational,
intentional agents unless there are major changes in the way our brain
works. Moreover, if you use the truth of pure mechanism as a premise in
deciding what to do, this premise will entail no particular moral, legal,
or political conclusions. It will provide no guide to how one should live,
including how one should respond to the truth of NAT.

Finally, the argument from common sense in favor of the justified belief
that we are conscious, intentional creatures is overwhelming. Consider
again, for example, the nature of law itself. As we have seen, law is a system
of rules that at the least is meant to guide or influence behavior and thus
to operate as a potential cause of behavior. In brief, it would be impossible
at present for us to abandon the well-justified belief that action may be
influenced by reason and that our intentions are causally efficacious.[22] Is it

---

[22]   As Jerry Fodor writes,

> if commonsense intentional psychology were really to collapse, that would be,
> beyond comparison, the greatest intellectual catastrophe in the history of our
> species; if we're that wrong about the mind, then that's the wrongest we've ever
> been about anything. The collapse of the supernatural, for example, doesn't
> compare . . . Nothing except, perhaps, our commonsense physics . . . comes as

an illusion for Professor Wegner to believe that he deserves the credit (and royalties) for *The Illusion of Conscious Will*, because it was really only his brain that wrote the book, and brains don't deserve credit and royalties?

The new neuroscience does not yet pose a threat to our fundamental conception of personhood and actions and all that follows from it, including the concept of responsibility and related concepts, such as mens rea. Indeed, many think that the most important task for the new neuroscience is to understand folk psychology, to explain how mental states are causally efficacious. At the very least, we remain entitled to presume that conscious intentions are causal and to place the burden of persuasion at a very high level on proponents of NAT—a burden that at present is not close to being met.

The Challenge of Determinism

For reasons similar to those just adduced, the new neuroscience casts little doubt on responsibility generally. People often think that the discovery of causes of behavior over which people have no control suggests that determinism or universal causation is true, at least for caused behavior, and undermines "free will," which in turn is thought to be a precondition for responsibility and other worthy goods, such as human dignity. This thought is what terrifies people about the scientific understanding of human behavior, which relentlessly exposes the numerous causal variables that seem to toss us about like small boats in a raging sea storm. They are afraid that science will demonstrate that we are nothing but mechanisms. To see why this view is mistaken, however, and why the new neuroscience does not threaten responsibility generally, requires a brief general discussion about the concept of free will.

---

near our cognitive core as intentional explanation does. We'll be in deep, deep trouble if we have to give it up . . . But be of good cheer; everything is going to be all right.

Psychosemantics: The Problem of Meaning in the Philosophy of Mind, p. xii (Cambridge: MIT Press, 1987). The entire book is a defense of commonsense intentional explanation.

Before discussing the threat to free will generally, however, it is first important to understand that neuroscientific and other biological causes pose no more challenge to responsibility than nonbiological and social causes. As a conceptual matter, we do not necessarily have more control over social causal variables than over biological causal variables. In a world of universal causation or determinism, causal mechanisms are indistinguishable in this respect, and biological causation creates no greater threat to our life hopes than social causation.[23] For purposes of the free will debate, a cause is just a cause, whether it is biological, psychological, sociological, or astrological.

There is no uncontroversial definition of determinism, and we will never be able to confirm whether it is true. As a working definition, however, let us assume, roughly, that all events have causes that operate according to the physical laws of the universe and that were themselves caused by those same laws operating on prior states of the universe in a continuous thread of causation going back to the first state. Even if this is too strong, the universe appears so sufficiently regular and lawful that it seems we must adopt the hypothesis that universal causation is approximately correct.[24] If this is true, the people we are and the actions we perform have been caused by a chain of causation over which we had no control and for which we could not possibly be responsible. How would responsibility be possible for action or anything else in such a universe?

No analysis of this problem could conceivably persuade everyone. There are no decisive, analytically incontrovertible arguments to resolve the metaphysical question of the relation between determinism, free will, and responsibility. And the question is metaphysical, not scientific. Indeed, the debate is so fraught that even theorists who adopt the same general approach to the metaphysical challenge substantially disagree. Nevertheless, the view one adopts has profound consequences for moral and legal theory and practice. After describing the debate, I shall

---

[23] See Janet Radcliffe Richards, *Human Nature After Darwin: A Philosophical Introduction* (London: Routledge, 2000) (complete analysis of the indistinguishability of biological and social causation as threats to personhood and ordinary responsibility).

[24] Galen Strawson, see note 15, p. 12 (terming this hypothesis the "realism constraint").

consider why even a thoroughly material, causal view of human behavior does not undermine ordinary responsibility.

The Metaphysics of "Free Will v. Determinism"[25]

The first standard answer to the claim that the universe is causally homogeneous, "incompatibilism," is that mechanism is inconsistent with responsibility, and comes in two forms, hard determinism and metaphysical libertarianism. The former admits that mechanism is true—that human beings, like the rest of the universe, are entirely subject to and produced by the physical states and laws of the universe—and therefore claims that no one is responsible for anything. The latter denies that mechanism is a true explanation of human action, and that therefore responsibility is possible. The second standard answer, "compatibilism," also embraces mechanism but claims that responsibility is nonetheless possible. Some philosophers, such as Daniel Dennett, believe that the debate should be abandoned and that responsibility can be justified without taking a position.[26] I believe, however, that although the debate is unresolvable, considering responsibility requires taking a position and that writers such as Dennett have not avoided the debate.

I will consider the three positions in order, oversimplifying massively but not blurring the broad outlines and central issues. In brief, I suggest that hard determinism can neither explain our practices nor ground a theory of desert and that libertarianism is metaphysically implausible and also cannot explain our law. I argue that only compatibilism can explain and justify our moral and legal responsibility doctrines and practices, including the uniqueness of human action, and that it insulates responsibility from the challenges that scientific understanding seems to pose.

Hard determinists and libertarians agree that "real" or "ultimate" responsibility is possible only if human beings have contra-causal

---

[25] The material in this subsection may be rather dense for those without a background or taste for philosophical argument. If one thinks that a causal physical view of the universe undermines the possibility of responsibility, however, this type of analysis cannot be avoided.

[26] See Daniel Dennett, *Freedom Evolves*, pp. 1–22 (New York: Viking, 2003).

freedom—that is, the freedom to originate action uncaused by prior events and influences. This is what is typically meant by "free will." Otherwise, both agree, human beings have no real freedom, even if reasons are undeniably part of the causal chain that leads to action. Contra-causal freedom also appears to be the type of freedom that most ordinary people intuit or perceive that they in fact possess and that makes them responsible. Incompatibilists disagree, of course, about whether we have contra-causal freedom. Hard determinists believe that determinism or mechanism is true, that we lack contra-causal power, that acting for reasons is ultimately as mechanistic as other causes, and therefore we are not responsible. Eliminative materialists, for example, are classic hard determinists. Metaphysical libertarians believe that human beings are unique because we do have contra-causal power—at least normal adults do—and that acting for reasons originates uncaused causal chains. This godlike power of agent origination is the libertarian foundation of responsibility for action.

Hard determinism generates an external, rather than an internal, critique of responsibility; that is, hard determinism does not try either to explain or to justify our responsibility concepts and practices. It simply assumes that genuine responsibility is metaphysically unjustified. Even if an internally coherent account of responsibility and related practices can be given, it will be based on an illusion.[27] To see why hard determinism or mechanism cannot explain our responsibility attributions, remember that such causal accounts "go all the way down": determinism or mechanism applies to all people, to all events. Causation is not partial, and lack of knowledge about how something is caused does not mean that it is uncaused. Determinism is not a degree or continuum concept. To say, as many do, that there is a continuum between determinism and free will is simply a conceptual error.[28] Thus, if determinism is true and is

---

[27] See Saul Smilansky, *Free Will and Illusion*, pp. 40–73, 145–219 (Oxford: Oxford University Press, 2000) (arguing that free will is an illusion, but an illusion that is indispensable).

[28] See, for example, Committee on Addictions of the Group for the Advancement of Psychiatry, "Responsibility and Choice in Addiction," 53 *Psychiatric Services* 707, 708 (2002). (example of conceptual confusion). What such claims usually mean is that responsibility is a continuum concept. If the capacity for rationality is the touchstone of responsibility, as the sections beginning on pages 159 and 172 suggest, then responsibility can be a continuum concept.

genuinely inconsistent with responsibility, then no one can ever be really responsible for anything and responsibility attributions cannot properly justify further attitudes or action. But Western theories of morality and the law do hold some people responsible and excuse others, and when we do excuse, it is not because there has been a little local determinism at work. Hard determinism can produce an internally coherent, forward-looking consequential system that treats human action specially and that might possibly encourage good behavior and discourage bad behavior. Nevertheless, hard determinism cannot tell us which goals we should pursue and it cannot explain or justify our present practices, which routinely pursue nonefficiency goals, such as giving people what they deserve, even if doing so is not the most efficient outcome.

The libertarian concedes that determinism or mechanism may account for most of the moving parts of the universe, but argues that human action is not fully subject to causal influence. The libertarian is thus able to distinguish action from all other phenomena and to ground responsibility in contra-causal freedom. If the libertarian is correct, the "buck" really does stop with human intentions, at least if intentional action is both rational and not influenced by coercive threats. The difficulty is that libertarianism produces a worthless view of responsibility or it depends on a "panicky" metaphysics, to use P. F. Strawson's phrase.[29]

One form of libertarianism holds that human actions are the product of indeterministic events in the brain, but why should such events ground responsibility? In what way are actions produced by indeterminate or random brain events "ours" or an exercise of a freedom worth wanting? If our brain states and the world in general are truly random or indeterministic, then it is difficult to imagine how responsibility could be possible. Such randomness is deeply implausible, however. We could not explain the regularity of physical events and of human interactions, nor could we explain the dependable relation between intentions and actions, unless there was a cosmic coincidence of astounding proportion

---

[29]   P. F. Strawson, "Freedom and Resentment," in *Free Will,* p. 80 (Gary Watson, ed., Oxford: Oxford University Press; 1982).

that accounted for the regularity. More important, brains would be akin to random-number generators, and our behavior would be equivalent to the random numbers generated. This is scarcely a secure foundation for responsibility or for any form of moral evaluation, because rational intentions would be random rather than the product of genuine practical reason. In a sense, nothing would be "up to us." In sum, if rational action is simply a product of biophysical indeterminacy, no one should be held responsible for any action.

An apparently more plausible version of libertarianism concedes that prior events and experiences affect our mental states but alleges that our actions are ultimately uncaused by anything other than ourselves. To begin, such a theory cannot specify the nature of the causal relation between an agent and an act that she causes.[30] Furthermore, there is no observational evidence that would confirm that agent causation is true. Our experience of psychological freedom when we act intentionally is often cited, but such arguments from experience for contra-causal freedom have been properly criticized on many grounds.[31] Perhaps most important, our experience might simply be a psychological fact about us rather than good evidence to justify the truth of agent origination. Moreover, why would we choose to adopt such an implausible theory when there are also no good nonobservational grounds, such as coherence or parsimony, for doing so?[32] Finally, our desire to believe that agent causation is true or the importance to our self-conception of this truth is not independently good reason to accept that truth if the metaphysical foundation is false. Metaphysical libertarianism is simply too extravagantly implausible to be a secure foundation for responsibility.

Compatibilism—which accepts the realism constraint—offers the most plausible causal metaphysics and the only coherent positive explanation of our current legal (and moral) practices. I also believe that a satisfactory normative defense of compatibilism is possible, based on

---

[30] See Michael S. Moore, *The Metaphysics of Causal Intervention*, 88 *California L. Rev.* 827 (2000).

[31] For example, John Stuart Mill, *An Examination of Sir William Hamilton's Philosophy*, in J. M. Robson, ed., *Collected Works of John Stuart Mill* IX pp. 449–457 (1979).

[32] Bok, see note 10, p. 45.

conceptions of personhood, dignity, and the like. For the purpose of this essay, however, which is concerned with the implications of one class of causal explanation within a thoroughly material worldview, a positive account is sufficient.

The form of compatibilism most consistent with the legal view of responsibility is not concerned with whether we have "ultimate" responsibility. Instead, it treats responsibility practices as human constructions concerning human action and asks if they are consistent with facts about human beings that we justifiably believe and moral theories that we justifiably endorse. Although legal cases and commentary are replete with talk about "free will" being the basis for responsibility, especially if a biological abnormality seemed to affect the legally relevant behavior in question, the presence or absence of contra-causal freedom or something like it is not part of the criteria for any legal doctrine that holds some people nonresponsible. If the realism constraint is true, all behavior is caused, but not all behavior is excused, because causation per se has nothing to do with responsibility. If causation negated responsibility, no one would be morally responsible, and holding people legally responsible would be extremely problematic. An assertion that "free will" was or was not present is simply a conclusory statement about responsibility that had to have been reached based on other criteria.

The Criteria for Responsibility

Virtually all legal responsibility and competence criteria depend on assessment of the agent's rational capacities in the context in question. For example, a person is competent to contract if he or she is capable of understanding the nature of the bargain; a person is criminally responsible if the agent was capable of knowing the nature of his or her conduct or the applicable law. To continue the example, some people who commit crimes under the influence of mental disorder are excused from responsibility because their rationality was compromised, not simply because mental disorder played a causal role in explaining the conduct. The rationality criterion for responsibility is perfectly consistent with the facts—most adults are capable of minimal rationality virtually all the time—and with moral theories concerning fairness and justice that we have good reason to accept.

Coercion or compulsion criteria for nonresponsibility also exist in civil and criminal law, although they much less frequently provide an excusing condition. Properly understood, coercion obtains when the agent is placed through no fault of her own in a threatening "hard choice" situation from which she cannot readily escape and in which she yields to the threat. The classic example in criminal law is the excuse of duress, which requires that the criminal must be threatened with death or serious bodily harm unless she commits the crime and that a person of "reasonable firmness" would have yielded to the threat. The agent has surely acted intentionally and rationally. Her completely understandable desire to avoid death or grievous bodily harm fully explains why she formed and executed the intention to commit the crime. The reason we excuse the coerced agent is not that determinism or causation is at work, for it always is. The genuine moral and legal justification is that requiring human beings not to yield to some threats is simply too much to ask of creatures like ourselves. Now, how hard the choice has to be is a moral, normative question that can vary across contexts. A compulsion excuse for crime might require a greater threat than a compulsion excuse for a contract. But in no case does compulsion have anything to do with the presence or absence of causation per se, contra-causal freedom, or "free will."

A persistent, vexed question is how to assess the responsibility of people who seem to be acting in response to some inner compulsion, or, in more ordinary language, seem to have trouble controlling themselves. Examples from psychopathology include impulse control disorders, addictions, and paraphilias (disorders of sexual desire). If people really have immense difficulty refraining from acting in certain ways through no fault of their own, this surely provides an appealing justification for mitigation or excuse. But what does it mean to say that an agent who is acting cannot control himself? I have explored this question at length elsewhere,[33] so I shall be brief and conclusory here. People who

---

[33] Stephen J. Morse, *Uncontrollable Urges and Irrational People*, 88 *Virginia LR* 1025 (2002).

act in response to such inner states as craving are intentional agents. A drug addict who seeks and uses to satisfy his or her craving does so intentionally. Once again, simply because an abnormal biological variable played a causal role—and neuroscientific evidence frequently confirms this[34]—does not per se mean the person could not control himself or had great difficulty doing so.

I believe that cases in which we want to say that a person cannot control himself and should be excused for that reason can be better explained on the basis of a rationality defect. In short, at certain times or under certain circumstances, the states of desire or the like make it supremely difficult for the agent to access reason. As always, causation and free will are not the issue. The assessment of human action in terms of rationality or commonsense criteria such as "self-control" is at issue. Lack of control can only be finally demonstrated behaviorally, by evaluating action. Although neuroscientific evidence may surely provide assistance in performing this evaluation, neuroscience could never tell us how much control ability is required for responsibility. That question is normative, moral, and, ultimately, legal.

In short, free will and causation are not criteria for responsibility or excuse. Therefore, in principle, no amount of increased causal understanding of behavior, from any form of science, threatens the law's notion of responsibility unless it shows definitively that we humans (or some subset of us) are not intentional, minimally rational creatures. And no information about biological or social causes can show this directly. It would have to be demonstrated behaviorally. Even if a person has undoubted brain abnormalities or suffered various forms of severe environmental deprivation, if he or she behaves minimally rationally, he or she will be responsible. Indeed, some abnormalities may make people "hyperresponsible." Consider, for example, a businessperson with hypomania who is especially energetic and sharp in some stages of his or her illness. Such a person is undoubtedly competent to contract in this state, even though the abnormality is playing a causal role in his or her behavior.

---

[34] For example, Rita Z. Goldstein and Nora D. Volkow, and Marc N. Potenza et al., see note 13.

It is of course true that many people continue mistakenly to believe that causation, especially abnormal causation, is per se an excusing condition, but this is quite simply an analytic error that I have called the fundamental psycholegal error. It leads people to try to create a new excuse every time an allegedly valid new "syndrome" is discovered that is thought to play a role in behavior. But syndromes and other causes do not have excusing force unless they sufficiently diminish rationality in the context in question. In that case, it is diminished rationality that is the excusing condition, not the presence of any particular type of cause.

To explore the ramifications of the foregoing analysis for the new neuroscience, let us consider the example of brain abnormalities and violent or aggressive conduct.[35] Accumulating evidence from increasingly sophisticated diagnostic techniques, such as functional and structural imaging, demonstrates that various abnormalities predispose people to engage in violent conduct. Such information may have both predictive and therapeutic implications, although mandatory screening and any form of involuntary intervention, no matter how benignly motivated, would raise serious civil liberties issues. I discuss the civil liberties issues below. It might also have the effects of deflecting attention from other, equally powerful and perhaps remediable social causes and of treating the problem as solely within the individual rather than as the product of an interaction between the individual and his or her environment. In the remainder of the discussion, however, I shall focus solely on the relevance of such information to personal responsibility.

The easiest case arises when the abnormality causes a state of severely altered consciousness. Criminal responsibility requires action and rationality, and in such instances the law holds either that the person did not act, because the definition of action requires reasonably intact consciousness, or that the action was not rational, because rationality requires the potential for self-reflection that altered consciousness undermines. Neuroscience evidence might well be relevant in such cases to assess the validity of the defendant's claim about his or her mental

---

[35] See, for example, Jeffrey L. Cummings and Michael S. Mega, *Neuropsychiatry and Behavioral Neuroscience*, pp. 360–370 (Oxford: Oxford University Press, 2003).

state. I address the admissibility of such evidence below. Of course, the law concerning the relevance of consciousness to responsibility was developed, and the law was able to evaluate such claims, before any of the modern neuroscientific investigative techniques were invented. Neuroscience thus teaches us nothing new morally or legally about these cases, but it may well help us adjudicate them more accurately.

The more problematic cases are those in which the defendant's consciousness was intact and he or she clearly acted, but an abnormality was present that may have played a causal role in the criminal conduct. As sophisticated people understand, abnormalities do not cause violent conduct directly and they are not excusing conditions per se simply because they played a causal role. Instead, they produce behavioral states or traits, such as rage, impulsiveness, or disinhibition generally, that predispose the agent to commit violent acts and that may be relevant to the agent's responsibility. After all, such states and traits can compromise rationality, making it more difficult for the agent to play by the rules. For example, younger children and people with developmental disability are not held fully responsible because it is recognized that their capacity for rationality is not fully developed. In these cases, once again, it is a rationality consideration, and not lack of free will, that is doing the work. Note, too, that if the capacity for rationality is compromised by nonbiological causes, such as child-rearing practices, the same analysis holds. There is nothing special about biological causation.

The law was cognizant of the relevance of diminished rationality to responsibility and developed its doctrines of mitigation and excuse long before modern neuroscience emerged. But unless neuroscience demonstrates that no one is capable of minimal rationality—a wildly implausible scenario—fundamental criteria for responsibility will be intact. On the other hand, neuroscience will surely discover much more about the types of conditions that can compromise rationality and thus may potentially lead to a broadening of current legal excusing and mitigating doctrines or to a widening of the class of people who can raise a legitimate claim under current law. Further, neuroscience may help adjudicate excusing and mitigating claims more accurately.

Informed Consent

It is a commonplace that the legal doctrine of informed consent protects a patient or research subject's liberty and autonomy. Competent adults have a right in virtually all circumstances to control what is done to or with their bodies and minds. There are controversies about how much information should be disclosed and what level of understanding a patient or research subject must achieve in order to make consent valid,[36] but all informed consent standards are based on the assumptions that the potential patient or subject is rational and that the information provided will aid the person's ability to make a rational decision about his or her self-determination. Once again, the law's model of the person as an intentional, rational agent grounds this doctrine. Exceptions to the need to obtain informed consent involve situations in which the person's autonomy interests are subjugated to other values, such as cases of compulsory treatment, or in which the person is not rational, in which case a substitute decision maker will be required.

Modern neuroscience raises at least two potential issues for the theory and practice of informed consent. The first is whether neuroscience can teach us anything new about the ability of people to process and to use information under various conditions, such as stress. Once again, the ultimate issue is behavioral—it is about a person's cognition rather than about the brain per se—but neuroscience will surely improve our understanding of information processing. Better understanding would be unlikely to alter the doctrine of informed consent profoundly unless it radically altered our model of the person. Indeed, most of the controversy about the requirements for informed consent and most legal developments have been produced by changing views about the moral issues, such as the degree to which autonomy must be protected and balanced against other values, and not by scientific data about the brain or behavior. Better understanding of cognition might alter practice considerably, however.

---

[36] See Jessica W. Berg, Paul S. Appelbaum, Charles W. Lidz, and Lisa S. Parker, *Informed Consent: Legal Theory and Clinical Practice*, 2d ed., pp. 41–74 (New York: Oxford University Press, 2001).

The second issue concerns consent to neuroscientific research. Doctrines of informed consent to research developed somewhat independently of and parallel with informed consent to treatment, but the justification is the same and is perhaps more important because being a research subject often brings no potential benefit to the subject other than altruistic satisfaction or some compensation, and it may impose substantial costs. Once again, improved understanding of brain function will not alter the fundamental legal doctrine unless the model of the person is changed, but neuroscience research does raise a number of important, interesting traditional issues.

Much research will be done on neurologically impaired people, which raises difficult informed consent issues if the impairments affect the potential subjects' rationality. Neuroscience may help to identify those incapable of giving adequate informed consent to neuroscientific and other forms of research. People with abnormalities are prone to the "therapeutic misconception," the error of believing that the research will benefit them, even when they are explicitly told that it will not or that it cannot be predicted that participation will help. This is a general problem about obtaining informed consent in a wide array of biomedical contexts. Finally, the complexity of the brain and its relation to behavior and to one's conception of the self raises somewhat speculative but profound issues. Biomedical research generally can potentially disclose threatening information, such as the presence of hitherto unrecognized disease or the potential for it. Neuroscience research can arguably discover information about the brain that could alter one's sense of self or that is especially invasive of the subject's privacy. Further, if neuroscientific investigation becomes more invasive, the potential for unpredictable effects on behavior and personality generally would increase substantially.

All of the informed consent to research issues just raised are traditional. I foresee no major changes in doctrine for neuroscientific research, but the application of existing doctrine may be contextually altered. For example, if some of the speculative problems arose, I assume that either especially rigorous informed consent would be required or that perhaps there would be state regulation. At present, however, the law and moral and political theory generally have the resources to deal as well with neuroscience research as with any other type of human subjects research with similar cost-benefit profiles.

Reform of Existing Legal Doctrines

This section first discusses generally under what conditions new neuroscientific discoveries (or other empirical advances in the understanding of behavior) might justify reform of legal doctrine. Then it turns to specific threats to civil liberties that neuroscience may present.

The Conditions for Legal Reform

The law is in many respects a conservative enterprise. Although it should be immediately apparent that economics has a great deal to teach the law, the critical reception of and controversy surrounding law and economics studies indicates that there will always be resistance to supposed reforms other disciplines suggest. For another example, despite the extraordinary advances in the understanding of mental disorder that have been made in the last half century and consistent calls for reform based on such understanding, the dominant version of the insanity defense is scarcely changed from the form adopted by the English in M'Naghten's case in 1843. In part, this conservatism may be warranted by controversy within the legally relevant disciplines, but it also derives from the failure of the data from other disciplines to undermine the behavioral assumptions upon which legal doctrine is apparently based or from the refractory nature of the moral or political values particular laws instantiate.

To return to the example of the insanity defense, and as should be apparent from the discussion of responsibility above, criminal law has long recognized that some criminals are incapable of basic rationality and has provided a defense to excuse such offenders. Advances in mental health science can teach us much about why some people lack rationality, and can help identify and treat those people, but it cannot tell society which rationality defects are sufficient to excuse a wrongdoer.

Deciding who is blameworthy and deserves to be punished is a moral and ultimately political question about which mental health science must fall silent. Moreover, mental health science cannot provide data that would relevantly undermine the behavioral assumptions grounding the insanity defense. Indeed, the law and mental health science share the identical behavioral assumption that some people are incapable of rationality. For a final example, economic analysis may demonstrate that certain levels of criminal punishment do not contribute marginally to consequential goals such as deterrence and that such additional punishment is therefore inefficient, but if people believe that some offenders morally deserve harsher punishments, efficiency will be jettisoned. Science cannot resolve the balance between deterrence and desert, which may often conflict.

In sum, before legislators and judges will be rationally justified in changing existing legal doctrine in response to the discoveries of any other discipline, at the least they should be convinced, first, that the data are valid; second, that the data are genuinely relevant to a particular doctrine; third, that the data convincingly imply specific changes; and fourth, that those changes will not infringe other values that may be more important. This is a tall order.

I am not an expert in all areas of law, but I am not aware of any major proposed or implemented doctrinal changes in any area of law based on the new neuroscience, including those areas, such as criminal law and mental health law, to which it appears most relevant. This is not surprising, for the reasons given in the earlier portion of this section and previous sections of this essay. Consequently, I can at most speculate about what may happen, based first on some analogous interdisciplinary work and then on the nature of neuroscience itself.

It appears that economics has had a substantial impact on doctrine and practice in those areas of law, such as antitrust, contracts, corporate law, securities regulation, and others, that deal primarily with economic activity and that primarily pursue the goal of efficiency. On the other hand, in many contexts there is resistance to economic analysis if efficiency conflicts with other goals, and newer research in what is termed behavioral law and economics is now casting doubt on many of the behavioral assumptions that grounded proposed doctrinal reforms. Nonetheless, when a discipline's theory and data are reasonably developed, the doctrinal implications are potentially clear, and consensually valid goals are not being undermined, doctrine will change in response to discoveries from other disciplines.

In recent years there has been an outpouring of scholarly work on law and the emotions, fueled by both the scientific and philosophical study of the emotions.[37] Although the bulk of this work has been addressed to criminal law issues, it does cast its net more widely. A central theme is that much legal doctrine is behaviorally unrealistic and unwise because it ignores the teachings of emotion theory. This scholarship is much newer than law and economics research and has not yet had much time to exert its influence, but I predict that it will have considerably less impact in the long run. In addition to the usual controversies about concepts and data, it is not clear that the law is unrealistic in the ways that critics in this school assert, that emotion theory has clear implications, and that there is much consensus about the goals to be achieved in those doctrinal areas to which emotion theory seems most relevant. Perhaps the concepts and database of emotion theory will advance sufficiently to produce more doctrinal impact.

Good neuroscience is hard science, but it will not avoid the usual controversies about theory, methods, and data. Assuming consensual validation, I predict that it will not have widespread, profound influence on doctrine in most areas unless, as I have suggested before, its discoveries radically alter our conception of ourselves. On the other hand, one can easily imagine substantial changes in discrete doctrines. For example, it is plausible to assume that neuroscience discoveries and investigative techniques might shed much light on topics such as memory for facial perception, which could cause changes in the law of evidence and police practices concerning eyewitness identification in criminal cases. Or perhaps neuroscience will be able to identify when people

---

[37] See, for example, Susan A. Bandes, ed., *The Passions of Law* (New York: New York University Press, 1999) (collection of representative essays from eminent lawyers and philosophers).

are consciously lying[38] or consciously or unconsciously discriminating on the basis of objectionable factors such as race.[39] Such discoveries could also have profound effects on evidentiary practices. But notice that even an exceptionally sensitive lie detection or discrimination detection technique might not be used because we fear state invasion of our innermost thoughts, even for undoubtedly worthy purposes such as discovering truth or uncovering conscious or unconscious discrimination. At present, there is considerable reason to doubt whether such findings are valid[40] and, a fortiori, whether these techniques would be suitable for legal use. But future neuroscience will surely make discoveries sufficiently valid for legal use in some contexts.

POTENTIAL THREATS TO CIVIL LIBERTIES: PRIVACY, PREDICTION, AND TREATMENT

Whatever effects neuroscientific discoveries have on discrete doctrines, they might well raise the specter of profound challenges to civil liberties that I will discuss under the rubrics of privacy, prediction, and treatment. Other sciences, too, might make discoveries that would raise similar challenges, so the following discussion surely generalizes. The potential of neuroscience to invade our privacy by revealing various aspects of our private, subjective experience may produce both the strongest reaction against its use and substantial regulation. On the other hand, the use of techniques that permit genuinely accurate lie detection and other valuable ends may be so alluring that the temptation to use them will be great. One need only think about our nation's legal response to the

---

[38] See, for example, D. D. Langleben et al., "Brain Activity During Simulated Deception: An Event-related Functional Magnetic Resonance Study," 10 *NeuroImage* 1 (2001).

[39] See, for example, E. A. Phelps et al., "Performance in Indirect Measure of Race Evaluation Predicts Amygdala Activation," 12 *Journal of Cognitive Neuroscience* 729 (2000); but see E. A. Phelps et al., "Intact Performance on an Indirect Measure of Race Bias Following Amygdala Damage," 41 *Neuropsychologia* 203 (2003).

[40] See J. T. Cacioppo et al., "Just Because You're Imaging the Brain Doesn't Mean You Can Stop Using Your Head: A Primer and Set of First Principles," 85 *Journal of Personality and Social Psychology* 650 (2003).

war on terror to recognize that justifying the use of privacy-invasive techniques may not be so difficult after all.

The question is what constitutional or legislative limits may be placed on such techniques. In a recent case that provides a technological analogy, for example, the Supreme Court held that the Fourth Amendment prohibition against unreasonable searches and seizures barred police use of heat sensors from outside a private home to detect marijuana plants within.[41] I think it is clear that the government will not be able to use neuroscientific investigative techniques to go on "mental fishing expeditions" generally, but various state interests may permit infringing hitherto protected interests. For example, the Supreme Court recently held that under limited conditions the state had the right to medicate with psychotropic medication a psychotic criminal defendant solely for the purpose of restoring the defendant's competence to stand trial.[42] The state's interest in adjudicating guilt and innocence in cases of serious crimes of violence and property was deemed sufficient to warrant infringing the defendant's admitted liberty interest in deciding whether to take such medication. Neuroscience undoubtedly poses a privacy threat.

Neuroscientific techniques might also increase the ability to make accurate predictions about various forms of future behavior. If some behaviors that are particularly socially problematic can be accurately predicted, once again there will be a temptation to use such techniques for screening and intervention, which would also pose a civil liberties threat. For example, criminal and antisocial conduct generally is an immense social problem in the United States. It is well known that over 2 million people are now incarcerated in state and federal prisons. Given the apparently strong association of various neuropsychological and neuropsychiatric abnormalities and some types of criminal conduct, and the increasing ability to detect such abnormalities, it is plausible to assume that neuroscientific techniques may well enhance the ability to predict future antisocial conduct among both those who have not

---

[41] *Kyllo v. U.S.*, 533 U.S. 27 (2001).

[42] *Sell v. U.S.*, 539 U.S. 166 (2003). The majority, although approving forcible medication to restore trial competence generally, thought that specific cases in which it would be justified would be rare. I suspect that there will be many cases in which forcible medication is approved by the lower courts. It will then remain to be seen whether the Supreme Court will place more specific limits on the practice.

yet engaged in such conduct and those who have. The social and personal costs of criminal conduct are so great that if the predictive techniques were sufficiently sensitive and remedial intervention of any sort were possible, once again there would be strong temptation to screen and intervene.

It would be far easier to justify screening among prisoners and others under criminal justice control. Although prisoners have rights, they may be severely curtailed, and techniques that increased the accuracy of recidivism predictions would probably be acceptable to promote public safety. On the other hand, widespread screening of apparently at-risk children and adolescents—even if the risk status were identified by objective, valid measures—would be legally and politically fraught, especially if the predictive technique and the necessary interventions were particularly invasive of liberty generally. Labeling and stigma effects and the potential for racial and ethnic bias would be terrifying. The widespread use of psychotropic medications such as methylphenidate (Ritalin) among public school children suggests, however, that a screening/intervention scenario would not be unthinkable if predictive accuracy and remedial intervention were sufficiently successful and the "side effects" of both could be strictly limited. At present, the science is not sufficiently advanced, political resistance would be intense, and it is probable that such schemes, even if adopted, would not survive a constitutional challenge. But it is difficult to envision how society would respond to techniques that identified risk-creating abnormalities highly accurately and to effective interventions that would prevent undoubtedly serious social and personal harms. Traditional attitudes toward privacy and liberty might be changed considerably.

The potential for direct biological intervention in the working of the brain and nervous system to change thoughts, feelings, and actions, often polemically characterized as the potential for "mind control," is particularly disquieting, and many consider this a greater threat to liberty than genetic intervention, which commands more public attention. The government already has the constitutional authority to compel the use of psychotropic medications under relatively limited circumstances, as the brief description above of the Sell case and the use of methylphenidate in the schools indicates, but the potential for widespread intervention to change behavior is apparent. As is well known, the biological and behavioral definitions of abnormality and disorder can be controversial. At the extremes, of course, there is little problem, but the criteria for abnormal brain structure or function are not obviously self-defining, the criteria for behavioral abnormality are even more fluid, and there is a tendency to pathologize problematic behaviors and the structures and functions that seem associated. Thus, there is no guarantee that a relatively value-neutral criterion of abnormality will impose strict limits on the ability of the state to compel behavior-altering interventions.

For example, the Supreme Court has decided that the state may involuntarily medicate a prisoner with psychotropic medication only if it is medically appropriate and necessary for the safety of the inmate or others in the prison. If these criteria are met, the prisoner's liberty interest in avoiding unwanted psychotropic medication must yield.[43] Although the provision of safe conditions in prison is an important state interest, there is widespread agreement that medication cannot be used solely to control prisoners' behavior. Consequently, the concept of "medical appropriateness" is doing the work. In the case of a manifestly psychotic and dangerous inmate who refuses to consent to treatment, there would be little disagreement about the appropriateness of involuntary medication. Now, however, mounting evidence suggests that a class of antidepressant drugs with a relatively benign side effect profile, the selective serotonin reuptake inhibitors, may reduce the incidence of violence among prisoners who do not obviously meet the DSM-IV-TR[44] criteria for a depressive disorder. It is extremely tempting to assume that many potentially violent prisoners have "underlying" or hidden depressive disorders or that the risk of violence is a pathology that is medically appropriate to treat. There is no incontrovertible conceptual

---

[43] *Washington v. Harper*, 494 U.S. 211 (1990). Older decisions from the state and federal courts have declared various institutional "behavior modification" programs impermissible violations of civil liberties, but in these cases the methods used were particularly egregious and they were clearly being used in a punitive manner for institutional control and not as forms of treatment. See, for example, *Knecht v. Gillman*, 488 F. 2d. 1136 (8th Cir. 1973).

[44] American Psychiatric Association, *Diagnostic and Statistical Manual of Mental Disorders*, 4th ed., text rev. (2000).

or empirical block to making such assumptions. It is therefore possible that courts might approve a program that compelled medication after appropriate screening in order to serve the goals of safety and "treatment."[45]

The example just given can of course be generalized. Once again, the state will have more power involuntarily to intervene in the lives of those it already controls, such as prisoners and soldiers, than in the lives of other citizens, but wider programs may be envisioned. Public health officials already pathologize violence, especially involving the use of guns, as a public health problem, and it is easy to imagine compelled treatment of the risk of violence as a justified method of protecting public health. Present involuntary outpatient commitment is generally limited to people with relatively serious mental disorders, but as the example above indicates, adroit redefinition of pathology and medical appropriateness might widen the state's net considerably. Again, the current science and the political will to accomplish effective widespread behavior control are lacking. Nonetheless, as screening and intervention methods become more precise and effective, there will be pressure to use them, and proponents will defend their constitutional legitimacy.

If neuroscience or other sciences ever reach the levels of understanding and efficacy necessary to make the foregoing civil liberties concerns a realistic possibility, it is difficult to predict what legislatures and courts will do. If there are pressing social problems that seem soluble by a technological fix, current political and constitutional constraints may weaken.

Enhancement of Normal Functions

The desirability and permissibility of laws permitting or even compelling access to enhancements of normal functions raise immensely difficult conceptual, moral, political, and economic questions.[46] In this section,

---

[45] See, for example, *State v. Randall*, 532 N.W. 2d 94, 106–10 (Wis. 1995) (the continued commitment of a person acquitted by reason of insanity who is no longer mentally ill is justified if the agent is dangerous, there is a "medical justification" to continue the commitment, and the commitment does not exceed the maximum term of imprisonment that could have been imposed for the crime charged).

I shall simply try to touch on the major issues. What is interesting once again, however, is that although new scientific discoveries may raise the stakes, the questions raised about justice, equality, liberty, and efficiency are thoroughly familiar, and rich theoretical resources already exist with which to address them.

Let us first make the controversial but plausible and necessarily simplifying assumption that we can identify a relatively value-free conception of normality and abnormality that will apply relatively uncontroversially to a wide array of cases. Unless such a conception is possible, it will be impossible to distinguish between treatment and enhancement, because that distinction is dependent upon a prior conception of normality/abnormality. The boundary between normality and abnormality can of course shift as conceptual understanding and empirical data advance, but if it makes sense to make the distinction, as I am sure it does, then a treatment/enhancement distinction will also have force.

The law already permits a wide array of enhancements for those who can afford them. Some are quite expensive—such as cosmetic surgery in the absence of disfigurement, psychotropic drugs prescribed to make people without a diagnosable disorder feel even better, and prep courses for standardized tests—and, consequently, their availability is limited to those with the resources to purchase them. Others, such as the use of caffeine or nicotine to enhance mental acuity, are quite inexpensive and thus available essentially to everyone. There is certainly no general presumption that enhancement is per se undesirable or immoral. The law regulates the sale and use of such enhancements very little or indirectly, by requiring warning labels, prescriptions, and other means that scarcely prevent access for those with the necessary resources. Private preference, conscience, and pocketbooks are thus the primary predictors of which people obtain which enhancements.

---

⁴⁶ See Allen Buchanan, Dan W. Brock, Norman Daniels, and Daniel Wikler, *From Chance to Choice: Genetics and Justice,* pp. 61–164, 181–203 (New York: Cambridge University Press, 2000). This volume is the most complete, sophisticated, balanced general treatment of these issues that I am aware of. Although it focuses on genetic interventions, the analysis is perfectly generalizable to enhancements from any source.

Some potentially enhancing agents are largely or entirely prohibited either generally, because the government has decided they are too dangerous for anyone to use them, such as certain stimulants, or in particular contexts, such as sporting events, in which their use is considered unfair or otherwise undesirable. Such limitations do not undermine the observation that enhancement by cognitive and biological techniques is already widely permissible and acceptable in our moral, political, and legal culture. This is not surprising in a society that values personal liberty and primarily uses market mechanisms to develop and distribute most goods.

The use of enhancements raises thorny questions of distributive justice when the enhancements substantially increase the possibility that the agent will thereby obtain other, socially desirable goods, such as access to better schools, jobs, or the like. Is it really fair, for example, that a student from a wealthy home who already has enormous educational advantages by going to better schools should have the additional advantage of taking a prep course for the SAT or of having access to a prescription for a substance that may increase her alertness, concentration, and other qualities that promote excellent performance on cognitive tasks? Many views of justice deny that this is fair, because they hold that most inequality is not justified, but others endorse the inequalities that result as justified by liberty, efficiency, and other values. We cannot resolve these issues, but we should note that as enhancements become more effective, the potential for unjust distribution will increase, especially if the original distribution of endowments and access to the enhancements are unfairly unequal.

The discoveries of neuroscience may well provide highly effective, precise enhancement possibilities that will affect physical and cognitive functions that strongly predispose to improved performance in important life tasks. Let us also assume that such enhancements would not have undesirable side effects. If so, and they are not freely available because they are too expensive for many citizens, then potentially unfair increases in inequality will result. This could be addressed by prohibition or by making the enhancements more freely available by subsidization or other mechanisms. The latter would not have the desired effect, however, because people do not have equal endowments to be enhanced. Unless, miraculously, an enhancement caused everyone to produce precisely the same performance, the whole performance distribution would simply shift upward, but the original inequalities would remain. We should also note that highly effective enhancements may be used in the service of vice and not just for virtuous pursuits.

Enhancing everyone in certain ways may perhaps be socially desirable and should be implemented (by various inducement mechanisms) even if it does not reduce inequalities. For example, if people with low normal intelligence could enhance their cognitive abilities, then they and all of society might be better off, but such people would not become the cognitive equals of those better endowed *ex ante* if the latter were also permitted to use the same enhancement. It is also interesting to contemplate whether, to pursue greater equality, certain enhancements would be permissible only for those people whose normal abilities were below some threshold and prohibited for those above that threshold. I assume that such a scheme would be held unconstitutional at present as a denial of both liberty and equal protection, but if certain inequalities threatened the social fabric, one can imagine a court upholding such a law.

Finally, may the state make enhancements obligatory? Some enhancements already are imposed. Public education or some equivalent is a requirement for all citizens because the state interest in promoting a citizenry capable of economic productivity and informed participation in the political process is extremely weighty. No liberty is absolutely protected, and any may be infringed if the government purpose for infringement is sufficiently strong. A balance must always be struck. Suppose, for example, there was widespread agreement that general social improvement in cognitive skills would be desirable for reasons similar to those justifying compulsory education. Why shouldn't everyone, and especially the least well endowed, be compelled to accept a new, neuroscientifically discovered, nonharmful enhancement for the good of the whole society? The liberty of those who did not wish to be enhanced would be infringed, but perhaps the infringement would be justified.

Consider the following analogy. Forced inoculation—and note that preventive inoculations are another form of enhancement—might be imposed on all citizens to avoid a dreadful infectious disease epidemic, including on those people who objected strongly on religious, moral, or other grounds. The examples are distinguishable, of course. One might say that failure of some people to enhance themselves cognitively does not threaten to make society worse off; it just fails to make some people better off. In contrast, failure to inoculate only threatens to make society worse off. The distinction is genuine, but the baselines against which welfare is assessed are normative and shift easily. It would not be difficult to reconceptualize refusal in the cognitive case as threatening harm. For example, a more communitarian society that expected citizens to exert their best efforts and to accept enhancements in order to increase the welfare of all would treat a person who refused to accept enhancement as a threat to the society. In sum, as the social benefits of an enhancement increase, the state interest in imposing it will also increase, but traditional concerns for liberty and freedom of thought and expression should politically and legally constrain compelled enhancement.

The widespread availability of effective enhancements could profoundly affect our conception of normality, raising the threshold of normality considerably. If this occurred, then certain abilities that were previously considered normal would be considered abnormal and would thus qualify for treatment, not enhancement. If this occurs and the disadvantage of those below the normality threshold is substantial, then such people would have a strong justice claim that the state should provide such treatment if they cannot afford it. According to virtually all current moral and political theory, the duty to provide treatment to the least well off is far greater than the duty, if any there be, to provide enhancement. But this is a topic for another paper.

The Admissibility of Neuroscientific Evidence

Many of the social and legal problems that neuroscientific discoveries potentially pose will be addressed by Congress or state legislatures. Legislatures are not bound by the rules of evidence, however, which apply only to adjudication, and thus they may be swayed by science of dubious validity. Courts will not overturn laws that are based on what many or even most experts would consider questionable science, because in our system legislation is presumed valid and will be upheld if it appears to have some rational basis. This standard is usually quite permissive unless the legislation affects fundamental rights. Courts will defer to legislative judgments. Moreover, in those infrequent instances in which a legislature may direct the use of certain techniques in furtherance of a legislative scheme, evidence based on those techniques will be admissible even if courts might not have accepted such evidence in the absence of clear legislative direction. For example, suppose the legislature adopted some form of civil commitment scheme based on predictions of dangerousness and then directed the appropriate officials to use a particular neuroscientific predictive technique. As long as there was some evidence of its validity, a court at a commitment hearing would be bound to admit evidence the technique produced even if it were not the best technique available or even very accurate at all.

Specific evidentiary rules govern the admissibility of scientific evidence at trial.[47] In our federal system of government, each state's rules of evidence govern trials within the state, and the Federal Rules of Evidence (FRE) govern in federal trials. Generally speaking, there are two dominant approaches to the threshold admissibility of scientific evidence: the *Frye* "general acceptance" standard,[48] and the Federal Rules standard as interpreted in *Daubert v. Merrell Dow Pharmaceuticals, Inc.*,[49] *General Electric Co. v. Joiner*,[50] and *Kumho Tire, Ltd. v. Carmichael*,[51] an interpretation now codified in the Federal Rules.[52] Neither standard has constitutional status but is simply a rule of evidence that may be

---

[47] David L. Faigman, David H. Kaye, Michael J. Saks, and Joseph Sanders, *Science in the Law: Standards, Statistics and Research Issues*, pp. 1–65 (Minnesota: West Group, 2002) (most comprehensive, easily accessible overview of the law of evidence governing the admission of scientific evidence).

[48] *Frye v. United States*, 293 F. 1013 (D.C. Cir. 1923).

[49] 509 U.S. 579 (1993).

[50] 522 U.S. 136 (1997).

[51] 526 U.S. 137 (1999).

[52] Federal Rules of Evidence 701 (Opinion Testimony by Lay Witnesses), 702 (Testimony by Experts), 703 (Bases of Opinion Testimony by Experts). See also Rule 402 (Relevant Evidence Generally Admissible).

changed legislatively and by judicial interpretation. The *Frye* standard, which was previously the majority rule, admits scientific evidence if the data or techniques used by the expert witness have gained "general acceptance" in the field in which they were developed. This test thus uses the relevant disciplinary community as the standard.

In *Daubert*, the Supreme Court, in its capacity as the ultimate supervisor of the rules that apply in federal cases, was asked to interpret the rule governing expert testimony, but its opinion was such a broad interpretation that the law was changed considerably. Instead of in essence delegating the admissibility decision to the relevant disciplinary community, *Daubert* required federal judges themselves to act as the "gatekeepers," to make an independent evaluation of the scientific validity of the basis of the proffered evidence. The Court provided four nonexclusive criteria that courts must use to evaluate the scientific evidence: testability (or falsifiability), error rate, peer review and publication, and general acceptance. These are the factors to be considered, but there is no algorithm for how they must be considered. Moreover, even if proffered testimony appears scientifically valid generally, there must be a "valid scientific connection" to the legal question at issue. In other words, the scientific testimony must be relevant—it must assist the trier of fact to decide the question at hand. For example, valid evidence about the presence of an illness and its treatability might not assist the trier of fact if the question was what caused the condition. If the testimony is admissible, which is often decided at a preliminary hearing, then its weight, its persuasiveness, is for the jury or judge (in a "bench" trial) to decide. In most cases, appellate courts will overturn trial court admissibility rulings only if the trial court abused its discretion, which is a hard standard for the party appealing to meet.

The *Daubert* standard has been enormously influential among the states, and it is now the prevailing rule, although a minority of states cling to the "general acceptance" standard. *Daubert* is both more and less permissive. It is more permissive because it does not require general acceptance if the scientific foundation for the evidence is strong. On the other hand, it will not admit even generally accepted evidence unless the foundation is strong.

Daubert leaves open many questions that must be resolved either generally or in specific cases, such as what counts as a science, or what error rate is acceptable for what purposes. For example, consider a dangerousness prediction device that could be used to determine both whether a person should be civilly committed and whether he or she should be executed. Should the acceptable error rate be the same in both determinations? Despite leaving such questions unresolved, *Daubert* clearly sets the appropriate general standard—the strength of the scientific foundation. As a result, innovative, cutting-edge scientific discoveries are now routinely admitted under *Daubert* if the proponent of the evidence can demonstrate that it meets the test for scientific validity.

Current and future neuroscience evidence is or will be potentially relevant to an immense array of issues routinely decided in civil and criminal cases. Good neuroscience clearly is science and can be assessed using the *Daubert* criteria. If neuroscientific evidence is specifically relevant in an individual case, as *Daubert* requires, and it is based on competent science, it will be admitted.

Conclusion

The new neuroscience poses familiar moral, social, political, and legal challenges that can be addressed using equally familiar conceptual and theoretical tools. Discoveries that increase our understanding and control of human behavior may raise the stakes, but they don't change the game. Future discoveries may so radically alter the way we think about ourselves as persons and about the nature of human existence that massive shifts in moral, social, political, and legal concepts, practices, and institutions may ensue. For now, however, neuroscience poses no threat to ordinary notions of personhood and responsibility that undergird our morals, politics, and law.

Glossary

Anterior cingulate cortex: part of the human brain associated with pain and fear; also thought to be responsible for attention control and for detecting flaws and errors in thinking.

Apoptosis: naturally occurring cell death; it eliminates damaged or unnecessary cells, balancing the cell creation that occurs continuously in the human body.

Atomoxetine: a nonstimulant treatment for attention deficit disorder that is believed to work by altering the functioning of neurotransmitters, used by neurons to communicate.

Autonomic nervous system: the portion of the nervous system that regulates essential organ and systemic activity, such as cardiac and lung function and digestion.

Brain death: final and permanent cessation of activity in the central nervous system. Currently, brain death is the generally accepted definition of death. Unlike the cessation of the heartbeat—the classical definition of death—brain death can occur even if the heart is still beating and respiration is possible with the aid of a mechanical ventilator (see definition).

Brain plasticity: the ability of the brain to grow new neural connections and to modify or eliminate old ones in response to injury and experience.

Brachial plexus: a complex of nerves, formed by the cervical and first thoracic nerves, that serves the chest, neck, and arms.

Broca's area: an area of the human brain that is involved in some aspects of the production of speech.

Catecholamine: a class of organic compounds in the nervous system that includes epinephrine, norepinephrine, and dopamine; they act as hormones, neurotransmitters, or both.

Catechol-O-methyltransferase: an enzyme that deactivates catecholamines in the synapses (points of connection between neurons).

Cerebral angiogram: an X ray of blood vessels in the head, made visible by injection of a radioactive substance.

Cerebral cortex: the largest part of the brain. It consists of two sections (hemispheres), each containing four subsections (lobes): the

frontal lobe, the parietal lobe, the occipital lobe, and the temporal
lobe. It consititutes the "gray matter" of the brain.

Cholinergic system: the system of nerve cells that use acetylcholine,
a neurotransmitter that plays a major role at autonomic synapses and
neuromuscular junctions.

Dendritic: refers to a dendrite, the short protuberance from a neuron
that gathers the information sent across the synapse from neighboring
neurons and conducts it into the body of the cell.

Determinism: the philosophical doctrine that human action is
determined by external forces, rather than by free will. Also, in a
broader sense, a term used to describe the mechanistic nature of the
physical world in general.

Dorsolateral prefrontal cortex: part of the brain located in the front
of the cerebral cortex; it is involved in the human brain's "executive
functions," including deliberate actions, goal-directed behavior,
attention, planning, and decision making.

Dualism: any philosophical system that defines reality in terms of
irreducible principles, the most famous being mind-body dualism.
Although mind-body dualism can be traced back to Plato, the
philosopher René Descartes is considered its founding father. While
Descartes believed that the mind and body can interact, he saw
them as distinctly separate and composed of fundamentally different
substances. Thus, according to Descartes, the mind can exist without
the body, and the body without the mind.

Electroencephalograph (EEG): a device used to detect and record
brain waves.

Evoked potentials: measures of changes in the electrical activity of
the nervous system caused by external events or internal mental
processes. Exogenous evoked potentials are often used to diagnose
and monitor sensory pathways. They are classified according to the
stimulus used to evoke them. For example, brain-stem auditory
evoked potentials result from acoustic stimuli.

Free will: the philosophical theory that man has the ability to choose
a course of action from various alternatives. Free will is challenged
by determinism.

Frontal lobe: the single largest area of the human cerebral cortex; it is the site of emotions, cognition, and motion.

Guillain-Barré syndrome: caused by a misdirected immune response; Guillain-Barré syndrome is a disorder characterized by progressive paralysis, usually beginning in the legs. In most cases, complete or nearly complete recovery is possible.

Health Insurance Portability and Accountability Act (HIPAA): a law passed by Congress in 1996; it is designed to encourage electronic transactions within the health care industry, while also requiring new safeguards to protect the security and confidentiality of health information. The regulation covers health plans, health care clearinghouses, and those health care providers who conduct certain financial and administrative transactions electronically (for example, enrollment, billing, and eligibility verification).

HIPAA prohibits group health plans from using any health status–related factor as a basis for denying or limiting eligibility for coverage or for charging an individual more for coverage. The privacy standards, developed by the Department of Health and Human Services, provide patients with access to their medical records and increased control over how their personal health information is used and disclosed. They represent a uniform, federal minimum for privacy protections for consumers nationwide.

Hippocampus: located within the temporal lobe, the hippocampus is a brain structure involved in processing emotion, memory, and the sense of smell.

Indeterminism: the theory that some events, including human action, have no cause. Moral responsibility for these random events is therefore impossible, because of a lack of choice in outcome.

Insanity defense: a plea by the defense in a criminal case that the defendant was not responsible for his or her actions because he or she was insane at the time of the alleged offense. The rule of law defining "insanity" varies from state to state, but falls generally into the five categories listed below (or some combination thereof). Where a person is found to have been insane at the time of the commission of the act that would be a crime, that person is adjudged to be not guilty of having committed the crime (also referred to as "not guilty by reason of insanity"). The five general rules are:

— **American Law Institute (ALI) rule:** in drafting the Model Penal Code, the ALI developed a definition of insanity that has been adopted by many jurisdictions. The rule requires that, due to mental disease or defect, the person lacked substantial capacity to appreciate the criminality of his act, or to conform his actions to the law.

— **Durham rule:** in *Durham v. United States* (1954), the District of Columbia Appellate Court adopted the rule that the defendant is not criminally responsible if his act was the product of mental disease or defect. This "product" rule was widely criticized for creating problems in determining causation and is not followed in any state, although it has been adopted by statute in the Virgin Islands.

— **Irresistible impulse rule:** coupled with the M'Naghten rule in several jurisdictions, this rule provides for the effect of mental illness on the volitional component, or voluntariness, of the defendant's behavior. In other words, the defendant may not be criminally responsible if, due to mental illness, he cannot distinguish right from wrong, *or*, if he can make that distinction, he is incapable of choosing between them.

— **M'Naghten rule:** the M'Naghten rule stems from a nineteenth-century case in England in which Daniel M'Naghten shot the private secretary to Prime Minister Sir Robert Peel, mistaking him for the prime minister. M'Naghten was found not guilty by reason of insanity. The rationale for this finding, which became the M'Naghten rule of insanity, was that, due to a defect in reason caused by disease of the mind, the defendant did not know the nature of the act he was committing or, if he did, did not know that the act was wrong. The M'Naghten Rule was widely adopted among the states in the U.S. and is still followed in many jurisdictions.

— **Modified Federal Rule:** prior to the assassination attempt on President Reagan by John Hinckley, Jr., the federal courts used a rule that had both cognitive and volitional ("impulse") aspects. After Hinckley's successful insanity defense, Congress passed the Insanity Defense Reform Act (1984), which eliminated the volitional (impulse) prong of the rule within the federal system and placed causation entirely on the presence of mental disease or defect.

Intellectual property: unlike physical property, intellectual property is the product of invention and creativity and does not exist in a physical form. A body of law protects, licenses, and otherwise addresses intellectual property. In the United States, issues of intellectual property are primarily addressed by copyright, patent, trademark, or trade secret laws, or some combination thereof.

Kleptomania: a mental disorder whereby the individual has an irresistible impulse to steal regardless of potential economic gain.

Limbic system: an agglomerate of brain structures involved in the processing of emotions, learning and memory, and pain.

Madey v. Duke: suit brought by Madey, a former researcher and director of the Free Electron Lab at Duke University, against Duke for patent infringement. Madey brought his free-electron research to Duke in 1989, requiring substantial equipment. He holds two patents on research conducted using this equipment. Duke's alleged infringing activity took place after Madey's departure from the university in 1998.

Duke defended its actions as allowed by the "experimental use" exception to patent infringement. The federal district court granted summary judgment for the university. The court defined this "exception" as permitting uses that are entirely for academic, research, or experimental purposes, as opposed to commercial purposes. However, the Court of Appeals for the Federal Circuit narrowed that definition to include only uses that were for amusement or to satisfy idle curiosity, while expanding the concept of commercial purposes to include activities that were in furtherance of the defendant's legitimate business, if not overt commercial gain

(307 F.3d 1351 [Ct. App. Fed. Cir. 2002]). The appeals court then remanded the case for further proceedings. Duke filed a writ of certiorari with the Supreme Court but was denied.

The decision has raised alarms in the research community, where there are fears that the exemption for fair use of patented materials as pertains to research is being so limited as to be eliminated in all practical senses.

Magnetic resonance imaging:
— **MRI:** the common abbreviation for *magnetic resonance imaging*. This is a noninvasive radiological method that uses radiofrequency radiation and magnetic fields to construct three-dimensional structural images of the brain. Intravenous contrast agents can significantly enhance the images obtained.
— **fMRI:** the common abbreviation for *functional magnetic resonance imaging*. This refers to the use of MRI to measure the changes in blood flow during cognitive tasks, thus providing spatial and temporal information about brain activity.

Mechanical ventilation: use of a machine, called a ventilator, to assist with breathing.

Mnemonic function: the process of forming memories, involving the hippocampus and the neocortex (an upper region of the cerebral cortex unique to mammals). It is not clearly understood how memories are formed.

Neurology: the branch of science that deals with the study of the nervous system and its disorders.

Neuron: the main type of cell that is found in the nervous system, also commonly referred to as a nerve cell.

Neuroscience: the branch of the life sciences that studies the brain and nervous system. Among the areas of study included under the broadest definition are the physiology, chemistry, and molecular biology of the nervous system; issues of brain development; brain processes such as sensation, perception, learning, memory, and movement; and neurological and psychiatric disorders.

Neurotransmitter: a substance (such as acetylcholine) that transmits nerve impulses across the gap between neurons (the synapse).

Neurotrophic factors: also known as neurotrophins; a family of proteins that is responsible for the nutrition and maintenance of nervous tissue. Neurotrophins support and promote nerve growth. Nerve growth factor (NGF) is one such neurotrophic factor; it affects a variety of neurons in the body.

Norepinephrine: also commonly called noradrenaline. A type of neurotransmitter, it is released in the sympathetic nervous system (part of the autonomic nervous system).

Nucleus basalis: a group of structures in the brain and the upper brain stem that secretes acetylcholine, a neurotransmitter that modulates the excitability of distant neurons.

Occipital lobe/cortex: lobe of the brain, it contains the primary and secondary visual cortices.

Opacification: the introduction of a contrast medium to render a tissue or organ opaque for medical imaging purposes.

Orbitofrontal cortex: the region of the cerebral cortex that covers the basal surface of the frontal lobes of the human brain.

P300 brain wave: a specific electrical brain wave that is emitted in response to the recognition of an object or stimulus.

Parietal lobe/cortex: the lobe of the brain involved in perceiving stimuli such as temperature, touch, and pressure.

Patent: a type of property right granted to an inventor, giving the exclusive right to the use of some invention (including, for example, mechanical devices, chemical compounds, and manufacturing processes) for a limited period of time.

PET: common abbreviation for *positron-emission tomography*. This is a type of functional imaging that provides temporal and spatial information about changes in the brain by measuring changes in administered radioactive isotopes in different brain areas. This technique, which shows changes in blood flow during cognitive tasks, is similar to fMRI but more invasive.

Phenomenology: a philosophy founded by Edmund Husserl; it originated in the early 1900s and insists on separating philosophy from psychology. It holds that reality consists of intentional objects as they are understood in human consciousness and does not exist independently of human consciousness.

Physicalism: a theory that rests on the tenet that everything that exists is ultimately physical and that matter is the only reality.

Pontine auditory pathways: part of the pons, a portion of the brain stem that contains ascending and descending nerve tracts. The pontine auditory pathways transmit auditory information between various structures in the brain.

Prefrontal lobe/cortex: the most anterior lobe of the human frontal cortex; it is involved in emotion, complex thought, and problem solving.

SPECT: the common abbreviation for *single-photon-emission computed tomography*. Like PET, SPECT is a type of functional imaging that provides temporal and spatial information about changes in the brain by measuring changes in administered radioactive isotopes in different brain areas. This technique shows changes in blood flow during cognitive tasks.

SSRI: the common abbreviation for *selective serotonin reuptake inhibitor*. Neurotransmitters like serotonin are released by neurons to be taken up by nearby neurons. Neurotransmitters can also be reabsorbed by the neuron that originally released them, in a process called reuptake. SSRIs are a class of drugs that block the reuptake of the neurotransmitter serotonin from the synapse by the presynaptic nerve cell endings. By inhibiting reuptake, SSRIs leave higher levels of serotonin in the synapse, so that it is more readily available for uptake by other neurons.

Superior frontal gyrus: the convolutions of the outer surface of the frontal lobe of the brain occupying the superior position—that is, the position farthest from the feet.

Synapse: the physical space between two neurons. It is the area through which one excitable cell (neuron) communicates with the next.

Temporal lobe/cortex: lobe of the brain involved in some aspects of hearing, learning, and memory.

Thalamocortical: refers to the connections between the thalamus, a mass of nerve cells centrally located in the brain that serves to relay impulses to and from the cerebral cortex, and the cerebral cortex itself.

Theory of desert: a philosophical theory that explains the meaning and justification of deservingness. Typically, the concept of desert is connected to the actions of persons, and not to qualities that they may possess over which they have no control (as in inherited ability).

Tonsillar herniation: also known as brain herniation, this is the displacement of certain brain tissues and fluids outside the compartments in the head that they normally occupy. Such herniation often causes massive stroke and can rapidly result in death.

Transcranial Doppler sonography: a procedure that uses ultrasound to image the blood flow in the arteries located at the base of the brain. The test is used to detect blockage in the arteries that may be decreasing the flow of blood to the brain, as well as other changes in cerebral blood flow.

Ventromedial frontal lobe: a region of the human brain found within the frontal lobe that is associated with a variety of high-level cognitive tasks. There is evidence that damage to this region results in an impaired ability to think and reason in a social setting.

List of Participants

Judith C. Areen, Georgetown Law Center

Bicka Barlow, Defense Attorney

Erica Beecher-Monas, University of Arkansas at Little Rock
William H. Bowen School of Law

Floyd E. Bloom, AAAS Board of Directors, The Scripps Research
Institute

Joe S. Cecil, Federal Judicial Center

Ming W. Chin, California Supreme Court

Deborah Denno, Fordham Law School

Harold Edgar, Columbia University School of Law

Martha Farah, University of Pennsylvania

Mark S. Frankel, AAAS

Brent Garland, AAAS

Michael Gazzaniga, Dartmouth University

Steven P. Goldberg, Georgetown Law Center

Henry Greely, Stanford School of Law

Zach W. Hall, University of Southern California/Keck School
of Medicine

D. Brock Hornby, United States District Court for the District
of Maine

Owen D. Jones, Arizona State University College of Law and School of Life Sciences

Alan I. Leshner, AAAS

Stephen J. Morse, University of Pennsylvania Law School

Charles P. O'Brien, University of Pennsylvania School of Medicine

Haskell M. Pitluck (Retired), State of Illinois, Circuit Court

Adina Roskies, Massachusetts Institute of Technology

Barbara Jacobs Rothstein, Federal Judicial Center

Edward F. Rover, the Dana Foundation

William Safire, the Dana Foundation

Kristina M. Schaefer, AAAS

Laurence Tancredi, New York University School of Medicine

About the Authors

Mark S. Frankel, Ph.D., is director of the Program on Scientific Freedom, Responsibility and Law at the American Association for the Advancement of Science (AAAS), where he develops and manages the association's activities related to science, ethics, and law. He has directed AAAS projects on scientific misconduct, the use of scientific information in the courts, the policy implications of advances in genetics, and the effects of intellectual property law on access to science and technology. For more than a decade, he has served as the staff officer to the National Conference of Lawyers and Scientists, a joint committee of AAAS and the American Bar Association. He serves on the editorial boards of several professional journals and is a Fellow of AAAS. He was the director of the AAAS project on neuroscience and the law.

Brent Garland, M.S., J.D., is senior program associate in the Program on Scientific Freedom, Responsibility and Law at the American Association for the Advancement of Science (AAAS), an attorney, and an active member of the Virginia State Bar. He began his legal career as an assistant district attorney in North Carolina and is a former member of the faculty of the University of Virginia. His work is centered on biomedicine, bioethics, biotechnology, and issues arising from the intersection between science and law.

Michael S. Gazzaniga, Ph.D., is the David T. McLaughlin Distinguished Professor, the director of the Center for Cognitive Neuroscience, and the dean of the faculty at Dartmouth College. A graduate of Dartmouth and the California Institute of Technology, Dr. Gazzaniga has been appointed to the Board of Advisors of the Riken Brain Science Institute and the Board of Scientific Advisors of the Center for Neural Science at NYU. He is a presidential appointee to the MIT Board of Scientific Advisors, Brain and Cognitive Science Department, and serves on the President's Council on Bioethics convened by President George W. Bush.

Dr. Gazzaniga is the president of the Cognitive Neuroscience Institute, which he founded in 1982, and is the editor-in-chief emeritus of the *Journal of Cognitive Neuroscience*, which he also founded. He heads the McDonnell Summer Institute in Cognitive Neuroscience and is a Fellow of the American Academy of Arts and Sciences.

Henry T. "Hank" Greely, J.D., is the Deane F. and Kate Edelman Johnson Professor of Law and a professor, by courtesy, of genetics at Stanford University. He specializes in health law and policy and in legal and social issues arising from advances in the biosciences. He chairs the steering committee of the Stanford University Center for Biomedical Ethics, directs the Center for Law and the Biosciences, and codirects the Stanford Program on Genomics, Ethics, and Society. He is a member of the California Advisory Committee on Human Cloning and is chair of the ethics subcommittee of the North American Committee of the Human Genome Diversity Project. A graduate of Stanford and Yale Law School, Professor Greely clerked for Judge John Minor Wisdom of the United States Court of Appeals and for Justice Potter Stewart of the United States Supreme Court.

Stephen J. Morse, J.D., Ph.D., is Ferdinand Wakeman Hubbell Professor of Law and Professor of Psychology and Law in Psychiatry at the University of Pennsylvania. Trained in both law and psychology at Harvard, Professor Morse's expertise is in criminal and mental health law, with special emphasis on individual responsibility in criminal and civil law. A diplomate in forensic psychology of the American Board of Professional Psychology, a past president of Division 41 of the American Psychological Association (the American Psychology-Law Society), a recipient of the American Academy of Forensic Psychology's Distinguished Contribution Award, and a member of the MacArthur Foundation Research Network on Mental Health and Law, Professor

Morse is also a trustee of the Bazelon Center for Mental Health Law in Washington, D.C. Before joining the Penn faculty, Professor Morse was the Orrin B. Evans Professor of Law, Psychiatry and the Behavioral Sciences at the University of Southern California.

Megan S. Steven is a doctoral student in medical sciences at the University of Oxford. A 2002 graduate of Dartmouth College, where she majored in psychology with a minor in neuroscience, Ms. Steven was a recipient of a Rhodes Scholarship in 2002. She was the founding president of the Dartmouth chapter of the National Society of Collegiate Scholars and also founded an outreach program for disadvantaged New Hampshire middle school students.

Laurence R. Tancredi, M.D., J.D., is clinical professor of psychiatry at New York University School of Medicine. He has a private practice in New York City. A graduate of the University of Pennsylvania School of Medicine and Yale Law School, Dr. Tancredi received his psychiatric training from the Yale School of Medicine. The former Kraft Eidman Professor of Medicine and the Law and professor of psychiatry at the University of Texas Health Science Center in Houston, he is the author of numerous articles and several books on topics in law, ethics, and psychiatry. He currently serves on the Scientific Advisory Council of the American Foundation for Suicide Prevention, and the Board of Directors of the International Academy of Law and Mental Health.

Index